Animal-centric Care and Management

Animal-centric Care and Management

Enhancing Refinement in Biomedical Research

Edited by
Dorte Bratbo Sørensen, Sylvie Cloutier,
and Brianna N. Gaskill

CRC Press
Taylor & Francis Group
Boca Raton London New York

CRC Press is an imprint of the
Taylor & Francis Group, an **informa** business

First edition published 2021

by CRC Press
6000 Broken Sound Parkway NW, Suite 300, Boca Raton, FL 33487-2742

and by CRC Press
2 Park Square, Milton Park, Abingdon, Oxon, OX14 4RN

© 2021 Taylor & Francis Group, LLC

CRC Press is an imprint of Taylor & Francis Group, LLC

ISBN: 978-0-367-18102-4 (hbk)
ISBN: 978-0-367-18083-6 (pbk)
ISBN: 978-0-429-05954-4 (ebk)

Typeset in Times
by Deanta Global Publishing Services, Chennai, India

Contents

Preface

Since the publication of Russell and Burch's book in 1959, the concept of the 3Rs (*refinement, reduction,* and *replacement*) has been used as a framework for improving the welfare of laboratory animals. The concept is typically defined and used as follows:

Replacement means the substitution for conscious living higher animals of insentient material. Reduction means reduction in the numbers of animals used to obtain information of a given amount and precision. Refinement means any decrease in the incidence or severity of inhumane procedures applied to those animals which still have to be used.[1]

In brief, the 3Rs aim at reducing the suffering of conscious animals by decreasing the number used and the severity of the inhumane procedures. However, this view of the concept is overlooking Russell and Burch's statement that application of the 3Rs is more than reducing inhumanities:

We can therefore begin tentatively to think of a scale of well-being to distress ... In this way we are led to set our sights high in removing inhumanity, and to attempt always to drive the animal up to the highest possible point on the scale. Thus, we can aim at well-being rather than at mere absence of distress.[2]

They also acknowledged that daily husbandry procedures are an important part of the animals' lives and that focus on animal welfare should not be reserved only for the experimental procedures. However, the 3Rs are often inadvertently considered to be relevant only in the context of procedures and experimental interventions. One reason for this narrow interpretation of the 3Rs, could be that Russell and Burch named their book *The Principles of Humane Experimental Technique*, indirectly prompting the reader to focus on techniques rather than housing and handling. We believe it is time to gently push the "scientifically centered" view of the 3Rs to make room for an "animal-centered" view that puts the emphasis on the animals, their needs, preferences, and emotional states. By establishing an animal-centric view on laboratory animal housing and management, we aim to take one particular R, *refinement*, a step further, beyond Russell and Burch's most well-established idea of "elimination of inhumanities." Often – as stated by Russell and Burch themselves – *refinement* is promoted only to the level where the initiatives and actions to reduce suffering and increase animal welfare is not in any way a disadvantage to the scientific aims.[2] Sadly, *refinement* is also often severely limited by economic considerations.

Traditionally, *in vivo* studies are planned and designed with a focus on the phenomenon that is being modeled. For example, the planning of procedures for modeling *E. coli*-induced diarrhea in young pigs typically focus on inoculation and frequency of fecal sampling, ignoring how oral gavage and repeated restraint may add stress to the general experience for the animals. Selection of measures used to ensure animal welfare within the experimental setting is determined once the experimental procedures have been set. Perhaps the pigs will be given additional enrichment such as an empty cardboard box to manipulate and extra straw for rooting. In other words, the welfare of the pigs throughout the study is considered only within the limits and the framework set by the study design. What we are trying to accomplish with this book is to show that instead of fitting the animals into experimental conditions, one should first strive to adjust the baseline conditions (husbandry and housing) to better meet the needs and preferences of the animals, and then fit the experimental conditions within that framework. Thus, the first step when implementing an animal-centered strategy should be to determine the general approach to handling, housing, and husbandry. This first step will in most cases need to be implemented at facility management level to signal that these changes in handling and husbandry are prioritized by the management. Moreover, additional resources may be needed as animal-centric care and management in the start-up phase may be time consuming. For example, it may be necessary to train staff in new techniques (such as tunnel handling of mice or clicker training). In the example of the *E. coli* study in pigs, it could be that the chief executive of that particular animal facility decides – as a general rule – that in the future pigs should never be restrained for dosing or injections, but should instead be trained to cooperate for procedures using positive reinforcement techniques. Inoculums and oral medication should be administered in palatable food-items that the pig has learned to eat. All pigs must be comfortable with their caretakers and should willingly eat any food that is offered to them. Pigs could also be trained to enter an enclosure daily to get an identification number spray painted on their back rather than being ear tagged. After these factors have been determined, the researcher will have to fit their study design and methods into that framework. In other words, from the very beginning, each experiment is designed around making animal welfare the highest priority.

The higher the general expectations of animal welfare are, the more the animal's behavioral needs, preferences, and positive mental states are prioritized and defended. Even in studies where it is not possible to achieve a high level of animal welfare, if the baseline level is higher from the start then it is likely that some level of welfare will be maintained.

These alterations in thinking will require that we change the way we approach study design. This is likely to be a challenge, but our hope is that this book will provide ideas to help develop a more animal-centered approach to housing, handling, and not least, experimental procedures.

In this book, we have combined ideas and concepts on a variety of subjects from the fields of laboratory animal science, ethology, biology, and animal training to produce a laboratory animal science book that illuminates the uniqueness of the animals used for science, new ways to interact with them, and to provide stimulating and challenging environments. This book was contrived on the basic assumption that laboratory animals are all sentient, conscious, individuals with the capacity for both suffering and pleasure. Further, we use a broadly defined concept of "animal welfare;" we feel that a discussion on the nature of animal welfare is outside the scope of this book. It is our hope that this interdisciplinary book will act as a catalyst, resulting in multiple viewpoints and fields collaborating to optimize laboratory animal welfare.

It is also our intent that implementing an animal-centric approach to the husbandry and care of laboratory animals will raise the basic level of animal welfare. Additionally, this approach will fuel the development and implementation of housing conditions and handling procedures that will induce – to the best of our current abilities and knowledge – a long-term positive state of mind in the animals under our care. For this purpose, the present book includes species-specific chapters for animal species commonly used as experimental animals. Each chapter presents the newest and – to our minds – the most innovative and advanced ways of housing and caring for laboratory animals. Additionally, it discusses how to incorporate and execute positive reinforcement training both as cognitive enrichment as well as a way to enhance cooperation between animal and handler during experimental and husbandry procedures. The importance and nature of human-animal interactions, and suggestions on how to promote a culture of care, are presented. The nature of animal emotions, the importance of avoiding chronic stress, and the contribution of environmental factors to the development of abnormal behavior, are discussed.

This book may seem a little different because, in addition to building on peer-reviewed, statistically valid, high-quality research, we have also included anecdotes and experience-based knowledge. If a certain way of working with or training the animals has provided a positive effect over time, we have included this practice-based knowledge in the relevant chapters. By excluding this information, we would have risked losing valuable knowledge and experience, which could benefit the animals. However, we acknowledge that a single observation or anecdotes do not necessarily represent consistent, real, phenomena, but these observations are often the first step toward scientific inquiry. These observations fuel curiosity and provide insight into the formulation of a hypothesis. Thus, we encourage our readers to keep an open, inquisitive mind and accept the importance of experience and practice-based knowledge, because hypothesis-driven research may not yet have been conducted to evaluate it. Many experienced animal caretaker professionals (i.e., veterinarians, animal health technicians, caretakers, and some researchers) have developed intuition and gut feeling based on many years of informal observations and interactions with animals. This invaluable experience helps them notice subtle changes in their animals to assess their level of welfare. Hence, they are able to detect changes in their animals' affective states, even without using the few known indirect objective measurements (e.g., ultrasonic vocalizations in rats).

Furthermore, we acknowledge that anthropomorphic thinking is a valid starting point for creatively approaching challenges. However, such an approach should be mindfully combined with a thorough and detailed knowledge of the species' behavioral and ecological characteristics as well as the specific individual animal. Hence, we encourage developing animal care systems that are based around empathy and caring, as well as a meticulous knowledge of the animals and their unique behaviors. Although we may never fully measure or even understand what causes happiness or suffering in others, we must err on the side of caution and leave room for compassion and respect.

Many of us feel morally obligated to give animals a good life, especially if they are used to benefit humans. Hence, ensuring the welfare of animals is the primary motivation for establishing an animal-centric approach. All other added benefits from good animal welfare, such as more reproducible research data or higher animal caretaker work satisfaction, may be a side effect of this approach, but are not guaranteed and results may be variable. We therefore suggest aiming for a Culture of Care with the animal at its central focal point, since caring for the animals should be the highest priority. We also suggest aiming for a Culture of Courage. It can indeed be difficult to drive an animal-centric approach forward, and it may take a lot of courage to stand up against colleagues and facility managers focusing on high throughput and low costs. A Culture of Courage facilitates and motivates the individual person to put forward her or his ideas on how to continuously improve animal welfare despite the risk of subtle, "under the radar," resistance from colleagues. By explicitly stating this challenge, we hope that our colleagues sharing our animal-centric view will keep the faith as well as be inspired and motivated to promote and engage in animal-centric refinement actions in their own laboratories.

Moving outside one's own circles to discuss subjects such as animal welfare, care, and management with colleagues working in other disciplines or professions may spark good and fruitful discussions. Exchanging viewpoints and ideas on the care and management of animals with other researchers, veterinarians, animal caretakers, technicians, and animal trainers can be highly motivating. An increased awareness and recognition of each other's expertise, knowledge, and experience will promote creativity and enhance our ability to recognize new and better ways to improve enrichment. It may also enhance the quality and quantity of options available to the animal for displaying a wide range of species-specific behaviors. Researchers may have an extensive knowledge of animal perception and cognition, whereas animal caretakers know a lot about the actual behavior of the animal in various situations in the laboratory. Veterinarians may be able to ascertain possible pathological reasons for sudden changes in animal behavior, and animal trainers can train animals to cooperate with veterinarians and technicians. By listening to each other while approaching our common goal – the enhanced welfare of our animals – from each of our fields of expertise, we will be able to paint a bigger and more complete picture of the animal's experience. For this reason, the authors of this book come from various scientific fields, professions, and disciplines.

It is also our hope that this book will contribute to the debate on the obvious differences in care and husbandry that exist between animals within the same species but used for different purposes within the cultural boundaries of a society. Consider, for example, the differences between the care and husbandry of laboratory dogs vs. pet dogs, laboratory pigs vs. pigs raised for consumption, laboratory macaques vs. macaques kept in zoos, and zebrafish in the laboratory vs. a living room tank. The animals are the same; their emotional and their sensory capabilities are identical, and we should allow ourselves to be inspired by animal welfare initiatives and activities taking place in "the other worlds." It is the experience of some of the authors of this book that close collaborations and an open-minded exchange of experiences, thoughts, ideas, and animal management routines in laboratories, zoos, and companion animal fields will benefit everyone involved, especially the animals.

CONCLUSION

This book provides ideas to help develop an animal-centric culture and promote changes in management of laboratory species that are focused on the animals, their needs, and their welfare. Providing optimal welfare to laboratory animals should be the goal at all stages of the animal-based research process from shipping to the end of a study. It should be the responsibility of all of those involved in the process, not just the researchers or the veterinarians. It can be accomplished by setting high standards for the care, management, and housing of those animals, and assuring that the study fits within those criteria. Such an approach will be easier to implement if we change our mindset to focus primarily on the animals. We hope this book will be the first step towards making animal-centric management the "Gold Standard" for laboratory animal care, that it will emphasize increased awareness, and that it will promote an animal-centric culture in our laboratory animal facilities.

Dorte Bratbo Sørensen, Sylvie Cloutier, and Brianna N. Gaskill

REFERENCES

1. Russell, W. M. S. & Burch, R. L. in *The Principles of Humane Experimental Technique* Ch. 4, (Methuen & Co Ltd., 1959).
2. Russell, W. M. S. & Burch, R. L. in *The Principles of Humane Experimental Technique* Ch. 2, (Methuen & Co Ltd., 1959).

Editors

Chief editor, **Associate Professor Dorte Bratbo Sørensen** is based at the University of Copenhagen, where she teaches lab animal science. Her research interests revolve around the impact of different housing systems and various environmental enrichment or handling techniques on animal welfare and data quality. Another important area of interest is the implementation of training and socializing – especially the use of positive reinforcement training – as a way to enhance animal welfare and optimize the collection of physiological data. Together with Copenhagen Zoo, she arranges courses and seminars in positive reinforcement training and handling of laboratory animals, and she is the founder of the Centre for Laboratory Animal Training (CeLAT).

Co-editor **Dr. Sylvie Cloutier** currently works at the Canadian Council on Animal Care. Dr. Cloutier's research interest is on factors affecting the behavior and well-being of farm and laboratory animals, and the quality of human-animal interactions. She was a leader in introducing "rat tickling" as a method to improve handling of laboratory rats.

Co-editor **Dr. Brianna N. Gaskill** leads a research program focusing on welfare assessment of laboratory animals. She utilizes natural behavior, physiology, and affective state to assess an animal's overall well-being. She is especially interested in how better welfare can translate into better and more robust science. Her research interests include applied ethology, enrichment design and application, improving husbandry techniques, and how environment can affect scientific results when not tailored to the animal's needs and motivations.

List of Contributors

Jamie Ahloy-Dallaire
Department of Animal Science
Université Laval
Québec, Québec, Canada

Carolyn Allen
Department of Comparative Medicine
AbbVie
Lake County, Illinois

Robert E. (Bob) Bailey
Animal Behavior Enterprises LLC
Hot Springs, Arizona

Thomas Bertelsen
Novo Nordisk
Måløv, Denmark

Cathrine Juel Bundgaard
Novo Nordisk
Måløv, Denmark

Lori Burgess
Comparative Medicine Animal Resources
    Centre
McGill University
Montréal, Québec, Canada

Sylvie Cloutier
Assessment and Certification Program
Canadian Council on Animal Care
Ottawa, Ontario, Canada

María Díez-León
Department of Pathobiology and Population
    Sciences
Royal Veterinary College, University of
    London
Hatfield, United Kingdom

Isabel Fife-Cook
Department of Environmental Studies
New York University
New York, New York

Björn Forkman
Department of Veterinary and Animal Sciences
University of Copenhagen
Frederiksberg, Denmark

Becca Franks
Department of Environmental Studies
New York University
New York City, New York

Brianna N. Gaskill
Animal Sciences Department
Purdue University
West Lafayette, Indiana

Kelly Gouveia
Department of Biological Sciences
University of Chester
Chester, United Kingdom

Penny Hawkins
Research Animals Department
Royal Society for the Prevention of Cruelty to
    Animals (RSPCA)
Southwater, United Kingdom

Mette S. Herskin
Department of Animal Science
Aarhus University
Tjele, Denmark

Megan R. LaFollette
The North American 3Rs Collaborative
St. Louis, Missouri

I. Joanna Makowska
Applied Animal Biology
Animal Welfare Institute
University of British Columbia
Vancouver, British Columbia, Canada

Jeremy N. Marchant-Forde
Livestock Behavior Research Unit
United States Department of Agriculture
West Lafayette, Indiana

Jan Lund Ottesen
Novo Nordisk
Måløv, Denmark

Annette Pedersen
EAZA Animal Training Working Group
Copenhagen Zoo
Frederiksberg, Denmark

Andrea Polanco
Department of Integrative Biology
University of Guelph
Guelph, Ontario, Canada

Christine Powell
Central Animal Emergency Clinic
Coquitlam, British Columbia, Canada

Dorte Bratbo Sørensen
Department of Veterinary and Animal Sciences
University of Copenhagen
Frederiksberg, Denmark

Sarah Thurston
Unit for Laboratory Animal Medicine
University of Michigan
Ann Arbor, Michigan

Karolina Westlund
ILLIS Animal Behaviour Consulting
Hässelby, Sweden

1 Human-Animal Interactions

Megan R. LaFollette

CONTENTS

INTRODUCTION

Human-animal interactions are a daily, significant, and often unavoidable component of laboratory animal science that can impact research animals, the people who work with them, and even research outcomes.[1] After all, several regulatory bodies require that all animals are checked daily[2,3] and even this simple presence of a person in the housing room can influence the animals and research outcomes.[1] Unfortunately, without deliberate effort, human-animal interactions can negatively affect both animals and humans alike. Animals may experience fear and stress from these interactions – which can negatively affect their behavior, physiology, and quality of life. Humans may also experience stress from negative interactions or from performing stressful procedures such as euthanasia. Finally, scientific quality may suffer from negative outcomes from human-animal interactions; research results from stressed animals may be less likely to translate to human research – thereby decreasing study validity.

It is not surprising that most human-animal interactions in the laboratory are initially negative. Think of a mouse. In the wild, mice undoubtedly experience fear when encountering a human; we are much larger, and they likely view us as predators. Certainly, we are capable of causing them significant harm. In response, mice specifically try to avoid encountering humans by hiding or only entering human spaces in the dark, sticking to the periphery. If a mouse does meet a human, it may freeze and then immediately run away. But in the laboratory – even though the fear of humans remains – the mouse cannot flee or even hide. In response, the mouse experiences stress and may even try to bite the handler as their only defense to get away.

1

Many common laboratory species likely experience the same negative effects as mice. They also initially perceive humans as predators, therefore fearing even passive exposure or minor handling. In nature, these animals could cope with this fear by fleeing, hiding, or fighting. But in the laboratory, these coping mechanisms are hindered by the lack of space, hiding structures, and necessary, forced, close interactions. In addition to simple handling, laboratory animals are often exposed to common laboratory procedures that are inherently stressful or painful. For example, procedures such as restraint, blood collection, injections, and oral dosing can increase blood corticosterone, glucose, heart rate, and blood pressure[4] – all indicators of heightened stress. The stress from these procedures affects gene expression, behavior, and immune function[5,6] which can harm experimental validity and cause unwanted variability between animals. The negative impacts of handling and procedures have been recognized for more than 80 years.[7] However, these procedures often continue to be used without careful attempt to mitigate their negative impacts. Further, the consequences of stress from handling are often not accounted for in research studies.

Fortunately, human-animal interactions in the laboratory do not have to be negative (Figure 1.1). Instead they can be purposefully designed to benefit animals, humans, and research. When interactions are positive, they can lead to reduced stress in both animals and humans – making the experience enjoyable and enriching. Scientific research can be refined and scientific quality improved.[8,9] However, regardless of the benefits to humans and research, optimizing animal welfare should always be a priority. Providing laboratory animals with the best life possible is part of our ethical responsibility when engaging in animal research.

FIGURE 1.1   Positive human-animal interaction. A rat voluntarily seeks to interact with a human.

This chapter seeks to provide a broad-scale overview of human-animal interactions in the laboratory. What follows will address four main questions:

- What are human-animal interactions, and what types exist in the laboratory?
- How do these interactions affect laboratory animals and personnel?
- How should we interact with animals to improve human-animal interactions?
- What are the limitations to current research and potential avenues for future research?

DEFINING HUMAN-ANIMAL INTERACTIONS

Because human-animal interactions can vary substantially between studies, and even from day to day with an individual animal, clearly defining different human-animal interactions in the laboratory is necessary. Human-animal interactions can vary in duration (i.e., how long they last), depth (i.e., the intensity of the interaction), and valence (i.e., positive or negative quality of the interaction). Having clear, distinct, and consistent terminology for these interactions facilitates discussion and future research. A thorough review of terminology for human-animal interactions across fields already exists,[10] which will be briefly summarized here and supplemented with examples specific to the laboratory animal field (Figure 1.2).

The broadest and most widely applicable term, as demonstrated by its frequent usage in this chapter, is *human-animal interaction*. This term simply describes a sequence of behaviors that occurs between a human and an animal.[10] It applies whether the animals or humans recognize each other, interactions are repeated, or they are positive, negative, or neutral for either party. For example, this term is appropriate when a caretaker walks in a room to perform husbandry procedures, as well as when a researcher is habituating animals to specific research procedures. Overall, this term is the most commonly used and accepted across research publications and fields, in part because of its generality.

A slightly more restrictive term is *human-animal relationship*. This term generally refers to a series of interactions of any valence (i.e., positive, negative, or neutral) between animals and humans known to each other.[10] Humans may not recognize *individual* animals, but they must at least recognize the *group* of animals. Animals will typically recognize individual humans but may also recognize a group of humans (e.g., caretakers who wear a certain color of scrubs for

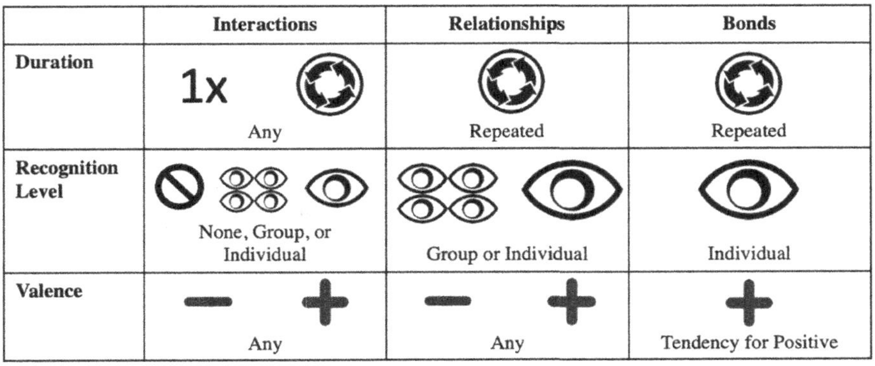

	Interactions	Relationships	Bonds
Duration	**1x** ↻ Any	↻ Repeated	↻ Repeated
Recognition Level	None, Group, or Individual	Group or Individual	Individual
Valence	**─ +** Any	**─ +** Any	**+** Tendency for Positive

FIGURE 1.2 Human-animal interaction terminology. A visual summary of the important distinctions between human-animal interaction terminology: human-animal interactions, human-animal relationships, and human-animal bonds. Duration indicates whether the interaction occurs a single time or is repeated. Recognition level indicates the level, if any, that both the human and animals recognize each other. Valence indicates whether the interaction(s) must be negative (−) or positive (+). Note that the depth and intensity of the interaction can vary for *all* definitions, although they are typically higher for human-animal bonds.

performing certain procedures). A human-animal relationship could exist between a room of rats in a particular study and their regular husbandry or animal health technician. The caretaker recognizes this particular group of rats when she works with them over time; simultaneously, the rats also likely recognize this caretaker. Thus, human-animal relationships occur frequently in the laboratory.

The final and most restrictive term is *human-animal bond*. Although there is some disagreement over its exact definition, there are three relevant and agreed upon elements.[11] First, a relationship must exist between a human and an *individual* animal. Second, the relationship must be reciprocal and persistent, meaning that *both* human and animal must recognize each other over a period of time. Third, interactions should tend to increase well-being for both parties. These bonds are most likely to form during long-term or small studies. They may also be more likely to form with animals with a closer evolutionary relationship to humans (e.g., non-human primates) or with companion animals such as dogs and cats.

All three terms can be used to describe different human and animal interactions in the laboratory environment. Human-animal bonds are likely the most beneficial human-animal interactions in the laboratory. However, human-animal interactions or relationships of a positive valence are also highly valuable. Consistently, using this terminology may increase understanding and communication of human-animal interactions within our own field and across different fields.

IMPACTS OF HUMAN-ANIMAL INTERACTIONS

In this section, three types of frameworks will be introduced and used to interpret the potential impacts of human-animal interactions for both animals and humans (Figure 1.3). First, an overall ecological framework will be discussed that can be used to describe interactions from either the animal or human perspective. Then an animal welfare framework will be introduced and used to discuss harms and benefits of laboratory animal research to animals. Finally, several theoretical frameworks will be introduced to discuss human-animal interactions from the human perspective. Both those frameworks and other research will then be used to discuss potential harms and benefits of laboratory animal research to humans.

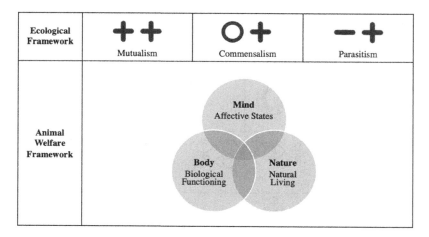

FIGURE 1.3 Frameworks for human-animal interaction. A visual representation of key frameworks from ecology and animal welfare that can be applied to human-animal interactions. For ecological frameworks, the "plus" indicates a benefit for one species, an open circle indicates no effect, and a "minus" indicates a harm for one species. The animal welfare framework[12] indicates three key areas of welfare and that the area in the middle (where all three elements overlap) results in superior animal well-being.

ECOLOGICAL FRAMEWORK

In ecology, interactions between two different species can be categorized purely on the harms or benefits to each (Figure 1.3). Using this framework can help us evaluate laboratory human-animal interactions with a broad lens. In *mutualism* both species benefit.[12] In *commensalism* one species benefits while the other experiences neutral impacts (neither benefits nor harms).[12] In *parasitism* one species benefits (often using the other as a resource) while the other is harmed.[12] In the laboratory, as well as in nature, interactions between species may not necessarily fit neatly into a single category and may change over time.[13]

Using this ecological framework, human-animal interactions in the laboratory can be classified into mutualism, commensalism, or parasitism – depending on the specific research topic and laboratory environment. As a whole, animal research has the potential to be mutualistic when humans benefit from increased scientific knowledge and animals benefit from animal-centric care, protecting them from predation, malnutrition, and disease. Mutualism may also occur if medical discoveries benefit both humans and animals, especially when the laboratory environment is well-managed to support animal welfare. Mutualism can also apply when animals are trained with positive reinforcement to cooperate with research procedures – this benefits the animal (as mental enrichment and welfare), the human's experience, and research data. On the opposite side of the scale, parasitism may apply when humans benefit from research while the utmost priority is not placed on animal welfare or animals receive harms from research procedures. For example, using animals in chronic variable stress models are, by design, stressful/harmful to animals but used for human benefit. Commensalism is likely rare, though it may occur in observational studies with moderate enrichment and little harm to animals.

ANIMALS

Human-animal interactions can also be discussed focusing on their harms and benefits to animals, especially through using a specific animal welfare theoretical framework.

Animal Welfare Theoretical Framework

Fraser et al. (1997) outlined a framework that evaluates welfare based on an animal's biological functioning, natural living, and affective states (Figure 1.3).[14] This framework can be used to assess the effects of human-animal interactions on animal welfare in the laboratory environment. Biological functioning is assessed by examining how interactions impact animal health and physiology. For example, standard laboratory routines such as restraint, injections, or oral gavage can increase corticosterone levels and decrease the immune function[4–6] – thereby decreasing welfare from the biological functioning perspective. Natural living is assessed by examining how interactions impact an animal's ability to express highly motivated natural behaviors. For example, rat tickling (a human-animal interaction that mimics aspects of rat social play) increases natural play behavior between pair-housed rats[15] – thereby improving welfare from the natural living perspective. Finally, affective states are assessed by examining how interactions affect the animal's emotions. For example, rat tickling increases positive emotions in laboratory rats[16] – thereby improving welfare from the affective states perspective. The affective states conception is considered the most vital element in animal welfare assessment because it addresses suffering or thriving. However, it is also the most difficult to measure. See Chapter 3 for a more in-depth discussion on animal emotions.

Harms for Animals

Overall potential harms of human-animal interactions for animals were outlined in the introduction of this chapter.

Benefits for Animals

Avoiding negative human-animal interactions and promoting positive interactions in the laboratory can improve welfare for many species in a variety of ways. For example, avoiding physical corrections (such as slapping, using a snout noose, or electric prod) in favor of scratching can increase pig growth rates, decrease corticosteroids, and improve reproduction.[17–19] In rats, using a modified, less restricted restraint technique for intraperitoneal dosing can decrease struggling, vocalizations, fecal count, and corticosterone.[20] In non-human primates, cooperative training programs can eliminate the need for chair restraint, increase animal welfare by allowing earlier detection of clinical symptoms, and improve the validity of scientific results by eliminating model confounds from stressful events.[21] Finally, exotic zoo animals (which share similarities to laboratory animals in long-term studies) can experience improvement to welfare when handled by fewer keepers and when spending more time with the familiar keepers.[22,23]

Human-animal interactions can also be used to intentionally induce positive emotions in laboratory animals through promoting play, conducting positive reinforcement training, and providing positive attention or touch. As explained above, play can be promoted in rats via rat tickling (Figure 1.4).[15,16] Its positive nature is particularly evident in studies showing its use as a reward in operant conditioning paradigms.[24,25] Positive reinforcement training itself is likely to engage the "seeking" circuit in the brain, which is considered to be highly rewarding and is associated with positive affect.[26] In dogs, positive human contact such as attention or stroking has been shown to decrease the stress response.[27] In cats, human petting increases positive affect and the production of secretory immunoglobulin A, which is beneficial for a healthy adaptive immune system.[28]

HUMANS

Human-animal interactions can also be described focusing on their harms and benefits to humans – several theories can be used to provide a framework for this discussion.

Human Theoretical Frameworks

Several theoretical frameworks have been proposed to explain human attraction to and benefits from animals. This section will present the three most common theories outlined by leaders in the field[29] and how they can apply to human-animal interactions in the laboratory.

FIGURE 1.4   Rat tickling. A rat receiving the positive human-animal interaction of rat tickling.

The "biophilia hypothesis" asserts that humans are genetically coded to respond to animals – which explains our innate attention and attraction to them.[30,31] In evolutionary history, there may have been genetic selection to be attentive to animals. For example, humans who were more attentive to other animals may have had greater fitness (higher survivability and reproduction) since paying attention to animals provides important cues about the environment.[32] In fact, research suggests that animals innately hold our attention and that we experience physiological changes in response to this attention. When comparing brain scans from people viewing landscapes, people, and animals, there is greater, category-specific neurological activation in the amygdala when viewing animals.[33] Animals can also provide a positive external focus of attention that may reduce cardiovascular responses to stress.[34] Viewing fish tanks can even effectively hold the attention of individuals with advanced Alzheimer's symptoms during meal times to help increase weight gain.[35] The biophilia hypothesis may also help explain our initial attraction to careers with animals, the reward we gain from positive interactions with laboratory animals, and why many of us find ourselves continually drawn to animals in our lives.

A second theory explaining human-animal interactions is the attachment theory. Attachment is a deep, enduring emotional relationship that connects two individuals and promotes a balance between physical closeness of the two individuals as well as independent exploration.[36,37] This theory was first proposed by Bowlby in 1958. The four characteristics of attachment (often seen in species that require care after birth) are proximity maintenance, safe haven, secure base, and separation distress. These features are adaptive for survival since they increase the likelihood that caretakers will provide resources to their young, or the individual under their care. Attachment is usually discussed in respect to parent-infant relationships, but could also apply to the human-animal bond.[38–41] We may see strong, secure attachments between humans and laboratory animals with close human-animal bonds. These laboratory animals may prefer to be close to their caretakers (especially in stressful situations), be more willing to explore in the presence of those caretakers and may even experience minor distress when separated from preferred caretakers. Their caretakers may be particularly responsive to these animals' needs, therefore promoting good animal care.

A third theory behind human motivation to interact with animals is the social support theory, which is the perception or reality that one is cared for, has access to supportive resources, and is part of a supportive social network – which includes animals.[29] This theory is often supported by experts in the field when exploring interactions between companion animals and humans[42] although results from pet ownership studies alone can be inconsistent depending on control factors.[42] Regardless, there is evidence that animals can buffer a stress response and promote positive physical and mental health outcomes.[43,44] This theory is more likely to apply with true mutualistic human-animal bonds in the laboratory where a caretaker feels they get support from their laboratory animals.

Harms for Humans

Of course, caretakers can experience harm from human-animal interactions. Simply performing or viewing stressful procedures (such as euthanasia) can lead to occupational stress or perpetration-induced traumatic stress.[45,46] It can be stressful to care for an animal that you will eventually euthanize; this is sometimes termed a "caring-killing paradox." [47,48] In one study, laboratory animal personnel who performed positive human-animal interactions (i.e., petting, naming, and talking to their laboratory animals) also reported higher secondary traumatic stress.[49] Furthermore, personnel may be subject to emotional harm from performing the "dirty job" of animal research that is often perceived negatively by society.[50] These factors could lead to compassion fatigue, comprised of burnout and secondary traumatic stress.[51] Having adequate social support (feeling like there is someone you can count on and talk to about your work with laboratory animals) is essential to account for the challenges of working in laboratory animal research.[49]

Human-animal bonds in particular may increase negative ethical and emotional implications of interactions in the laboratory. Once a bond is formed with a laboratory animal, ethical calculations around human-animal interactions may change. For example, most people feel greater responsibility to help and prevent harm to close friends versus strangers.[11,52] Therefore, laboratory animal caretakers may feel greater responsibility to the animals they are bonded to versus those they have a more distant relationship with. Severing this bond abruptly and without health-related reasoning (such as an end of study euthanasia of a healthy animal) may feel like a betrayal of trust[11] and lead to feelings of guilt. Therefore, establishing institutional support or discussion groups for personnel will likely be beneficial. When possible, another option is to allow an adoption policy for eligible, healthy animals to allow bonds to continue beyond the end of the animal's research career.

Negative scientific implications are also a possibility from human-animal bonds. These bonds could accentuate possible conflict of interest between a caretaker's allegiance to their animals versus the scientific research.[52] Having a bond with a certain animal could even lead to special care for that individual animal (e.g., providing extra treats or attention). Special care for one animal could cause unwanted variability in the response of that animal to experimental treatments or cause bias when assessing the animal during outcome evaluation. Overall, though, the potential harms from human-animal bonds can be mitigated with appropriate management – thereby allowing human-animal bonds to remain positive for all.

Benefits for Humans

Promoting positive human-animal interactions in the laboratory environment could benefit human psychological well-being, job efficiency, and professional quality of life. Psychological benefits could include reduced stress, anxiety, and more positive emotions at work. Anecdotally, people report feeling happier and more relaxed after tickling their laboratory rats or providing daily play to their laboratory cats. These innate feelings, combined with the knowledge of improving animal lives, could help mitigate moral distress from working with research animals and even promote compassion satisfaction. Furthermore, animal training and habituation could reduce the time required to perform various procedures (such as injections or restraint) and help these procedures be less stressful for handlers.[16,21] Higher quality of life, including reduced likelihood of burnout, among laboratory personnel is positively associated with providing more animal enrichment and engaging in more frequent positive human-animal interactions.[49] Overall, establishing a culture of care that includes positive human-animal interactions will likely be beneficial to caretakers and the research institution's culture (for more information on a culture of care, see Chapter 2).

Positive human-animal interactions are also likely to make laboratory animals better scientific research models. Specifically, animals that are positively handled may be less stressed, which can reduce variability in their response due to extraneous variables and may even increase reproducibility. For example, research with diabetic non-human primate models has shown that the utilization of positive reinforcement training enhances the value of translational research.[53] Better models are beneficial to researchers and to the general public, who both benefit from scientific research.

RECOMMENDATIONS

PRINCIPLES

From the time of the animal's arrival at a facility, it is crucial that human-animal interactions are positive (or at least neutral) and minimally stressful because these early contacts shape animals' expectations for future interactions with humans.[54] For example, animals should be carefully and gently handled when removed from shipping containers or separated from their mother after weaning. If possible, marking animals for identification should be done after a positive human-animal relationship has been established since marking often involves at least moderate restraint and sometimes distress.[15] Similarly, ensuring positive experiences between humans and young animals is particularly important, especially during their critical period of development,[55] although evidence for this suggestion is mainly supported by research with dogs and cats. Therefore, if laboratory animals are not bred or received at a particular facility until after this period, it may be beneficial to communicate with the vendor or provider about their protocols for sufficient positive human-animal interactions during this period. It is important to note that although neutral interactions do not elicit negative responses, they do not have as positive an impact as positive interactions.[56] Thus, one should aim for using positive interactions as much as possible.

In general, human-animal interactions should ideally improve an animal's control and predictability.[57] Studies in animal welfare show that the psychological aspects of predictability and control are highly important to both mental and physical health. For example, control and predictability influence the development of ulcers in rats undergoing electric shocks.[58] Predictability can be provided by either performing interactions at the same time each day or by providing a reliable and distinct cue before interactions – and this may be most important for aversive interactions. For example, cage changing can always be done in the same order by working from top left to bottom right. Additionally, training animals with positive reinforcement to cooperate with procedures can greatly increase predictability and control. Training allows animals to learn which interactions will result in rewards, choose whether to participate in interactions, and predict the interaction's outcome (for a further discussion see Chapters 5 and 6). Control in human-animal interactions in the home cage can be provided by allowing the animal choice in their location. For example, providing hiding structures can allow the animals to control whether they are viewed and reduces their fear of humans. Perhaps counter-intuitively, this can lead to more interaction. Zebrafish housed in enriched tanks with hiding structures are more likely to approach the front of the tank than those housed in barren ones (personal communication).

Familiarity with, and understanding of, species-specific behaviors can help promote positive human-animal interactions. Knowledge of species-specific behaviors can help caretakers determine what movements or actions of their own may appear threatening or aversive. For example, direct staring and eye contact can be threatening for primates, but support development of a human-animal bond with dogs. When in doubt, most species prefer movements that are slow, controlled, and predictable. Furthermore, knowing species-specific behaviors will ensure that caretakers know when they may need to intervene in self- or peer-directed behaviors. For example, rat social play appears very rough and could be misinterpreted as fighting by someone unfamiliar with the behavior. However, knowing that the focus of the contacts differs between aggression (rump) and play (nape) facilitates the recognition of the behaviors and helps determine if an intervention is necessary.

In addition to learning species-specific behaviors, caretakers should also become familiar with specific effective human-animal interactions with each species. For more detailed information, see each species-specific chapter in the second part of this book. In brief, here are some positive interactions that are recommended. Mice can be tunnel handled to reduce anxiety and increase control.[59,60] Rats can be tickled to decrease fear of humans, reduce stress, and increase positive affect.[16] Rabbits can be regularly, gently stroked from before six weeks of age through adulthood[61] and can be trained to cooperate with procedures (such as voluntary oral dosing rather than oral gavage)[62] to reduce stress. Dogs can be trained to cooperate with procedures, socialized, and petted or played with regularly.[27,63] Pigs can also be trained and scratched on the back.[64] Non-human primates can also be trained to cooperate during handling.[21,65] Zebrafish in enriched housing units which also have the ability to hide may benefit from calm, predictable attention such as slowly running fingers along the glass to promote schooling (personal communication). Each species likely can benefit from specialized, positive, human-animal interactions.

CHANGING HUMAN BEHAVIOR TO IMPROVE ANIMAL WELFARE

The laboratory animal science field has a responsibility to improve the lives of the animals under its care. Therefore, it is essential to select laboratory animal personnel that are committed to animal-centric care and provide them with sufficient training and support to provide the best care to laboratory animals. Individual contributions to human-animal interaction flourish best when supported by all levels of stakeholders, policies, and standards. Each individual contributes different knowledge and experience (e.g., program administrators, managers, clinical veterinarians, scientists, caretakers, students). Cohesive support for positive human-animal interactions will produce the best policies, ease implementation, and encourage compliance.

But what happens if some individuals are not convinced of the importance of human-animal interactions? Research with farm animal stock people shows that attempting to change a person's beliefs about human-animal interactions through education can help enact positive change[66] and there is preliminary evidence this could be the case for laboratory animals as well.[67] In one study, personnel were more likely to report implementing a positive human-animal interaction if they also reported stronger beliefs that the interaction was good, that their social and professional peers wanted them to do it, and, most importantly, that they had the ability to do it (e.g., time and education).[67] Therefore, management can first ensure that personnel feel that they have the ability to provide positive human-animal interactions by providing training on specific techniques and ensuring they are given adequate time to complete these tasks. During training, personnel should be instructed on both *how* and *why* to improve human-animal interactions. Training can instruct personnel on the benefits and importance of human-animal interactions, as well as outline the current social, public, and professional pressure for improved human-animal interactions in the laboratory.

LIMITATIONS AND FUTURE WORK

Of course, there are limits to our current knowledge and application of human-animal interaction in the laboratory. There is comparatively little research about the impact of human-animal interactions on research model health, affective states, and, ultimately, research validity or reproducibility. Furthermore, the specific details of human-animal interactions (i.e., habituation procedures, cage changing handling and frequency, handling for experimental procedures, daily and health checks) are often not reported in peer-reviewed publications; articles may simply indicate that the animals were "habituated" which does not provide enough detail for replication. This lack of research and reporting promotes skepticism in regard to the importance of positive human-animal interactions.

Some individuals are even concerned about potential harms to their research models. Until knowledge is increased through research and reporting, it will be difficult to mitigate such concerns. For example, using rat tickling could potentially change a depression model by encouraging positive affect and reducing typical signs of depression – therefore making the model ineffective. However, if the interactions between humans and animals for husbandry or experimental procedures have a notable impact on research outcomes, that may call into question the robustness, validity, and translatability of this model. Therefore, the impact of human-animal interactions on various research models should be systematically evaluated and reported.

Providing science-based evidence of the benefits of positive human-animal interactions will encourage their widespread integration into standards, guidelines, and procedures. It may also contribute to the development of new, more robust models. For example, rat tickling has been used to selectively breed rats that show an autistic-like phenotype.[68]

Another major and often cited limitation to implementing positive human-animal interactions is lack of time. Planning, learning proper techniques, and actually performing positive human-animal interactions can take a significant amount of time, resulting in extra personnel expenses. Very often, personnel working with, and caring for, animals already feel overworked and overwhelmed with their current tasks. Thus, it does not seem realistic to add to their already busy schedule. Although these are realistic barriers, every attempt should be made to make positive human-animal interactions a priority in animal laboratories and facilities. Keep in mind that changes do not have to be big and drastic. Often, even simple and relatively quick interventions can substantially change the human-animal relationship. The time needed to perform these interventions can be reduced as personnel acquire more experience in providing positive human-animal interactions. In addition, an animal responding to a positive intervention is more likely to approach and cooperate than attempt to escape and struggle. This may also contribute to reducing the time taken to perform procedures.

Furthermore, systematic investigations have shown that significant, positive human-animal interactions can be gained in relatively short amounts of time. For example, just 15 seconds of rat tickling for three days[15] and just five minutes of training dogs for oral gavage for four days[69] are beneficial. Although it is important that all staff contribute to these efforts, having a behavior and enrichment specialist may be useful to help coordinate and implement positive human-animal interactions.

CONCLUSION

In conclusion, human-animal interactions are just one aspect of animal-centric management, but a crucial one due to the vast number of interactions that are an integral part of animal-based research. Promoting positive human-animal interactions refines animal research by reducing pain, fear, and suffering – ideally promoting positive welfare. These interactions have the potential to improve research quality and personnel quality of life. When possible, fostering true, mutually beneficial human-animal bonds may be most advantageous. Overall, valuing and taking concrete steps to promote these unique relationships between laboratory animal personnel and their laboratory animals can ultimately improve both human and animal welfare.

REFERENCES

1. Cloutier, S. & Newberry, R. C. Physiological and behavioural responses of laboratory rats housed at different tier levels and levels of visual contact with conspecifics and humans. *Applied Animal Behaviour Science* **125**, 69–79 (2010).
2. Canadian Council on Animal Care (ed.). *Canadian Council on Animal Care Guidelines: Husbandry of Animals in Science* (Ottawa: Canadian Council on Animal Care, 2017).
3. National Research Council. *Guide for the Care and Use of Laboratory Animals* (Washington, DC: National Academies Press, 2011).
4. Balcombe, J. P., Barnard, N. D. & Sandusky, C. Laboratory routines cause animal stress. *Contemporary Topics in Laboratory Animal Science* **43**, 42–51 (2004).
5. He, Y. D. et al. Common handling procedures conducted in preclinical safety studies result in minimal hepatic gene expression changes in Sprague-Dawley rats. *PLoS ONE* **9**, e88750 (2014).
6. Issam, N., Abdelkrim, T., Ibtissem, C. & Narjess, K. Laboratory environment and bio-medical experience: The impact of administration technique on the quality of immune-behavior data results in stress experience. *Bioimpacts* **5**, 169–176 (2015).
7. Barrett, A. M. & Stockham, M. A. The effect of housing conditions and simple experimental procedures upon the corticosterone level in the plasma of rats. *Journal of Endocrinology* **26**, 97–105 (1963).
8. Russell, W. M. S., Burch, R. L. & Hume, C. W. *The Principles of Humane Experimental Technique*. Vol. 238 (London: Methuen, 1959).
9. Reinhardt, V. Compassion for animals in the laboratory: Impairment or refinement of research methodology? *Journal of Applied Animal Welfare Science* **6**, 123–130 (2003).
10. Hosey, G. & Melfi, V. Human-animal interactions, relationships and bonds: A review and analysis of the literature. *International Journal of Comparative Psychology* **27**, 117–142 (2014).
11. Russow, L.-M. Ethical implications of the human-animal bond in the laboratory. *ILAR J* **43**, 33–37 (2002).
12. Reece, J. B. et al. *Campbell Biology* (San Francisco, CA: Benjamin Cummings, 2010).
13. Leung, T. L. F. & Poulin, R. Parasitism, commensalism, and mutualism: Exploring the many shades of symbioses. *Vie et Milieu* **58**, 107 (2008).
14. Fraser, D., Weary, D. M., Pajor, E. A. & Milligan, B. N. A scientific conception of animal welfare that reflects ethical concerns. *Animal Welfare* **6**, 187–205 (1997).
15. LaFollette, M. R., O'Haire, M. E., Cloutier, S. & Gaskill, B. N. Practical rat tickling: Determining an efficient and effective dosage of heterospecific play. *Applied Animal Behaviour Science* **208**, 82–91 (2018).
16. LaFollette, M. R., O'Haire, M. E., Cloutier, S., Blankenberger, W. B. & Gaskill, B. N. Rat tickling: A systematic review of applications, outcomes, and moderators. *PLoS ONE* **12**, e0175320 (2017).
17. Hemsworth, P. H., Barnett, J. L. & Hansen, C. The influence of inconsistent handling by humans on the behaviour, growth and corticosteroids of young pigs. *Applied Animal Behaviour Science* **17**, 245–252 (1987).

18. Hemsworth, P. H., Barnett, J. L. & Hansen, C. The influence of handling by humans on the behaviour, reproduction and corticosteroids of male and female pigs. *Applied Animal Behaviour Science* **15**, 303–314 (1986).
19. Hemsworth, P. H., Barnett, J. L. & Hansen, C. The influence of handling by humans on the behavior, growth, and corticosteroids in the juvenile female pig. *Hormones and Behavior* **15**, 396–403 (1981).
20. Stuart, S. A. & Robinson, E. S. J. Reducing the stress of drug administration: Implications for the 3Rs. *Scientific Reports* **5**, 14288 (2015).
21. Graham, M. L. et al. Successful implementation of cooperative handling eliminates the need for restraint in a complex non-human primate disease model. *Journal of Medical Primatology* **41**, 89–106 (2012).
22. Mellen, J. D. Factors influencing reproductive success in small captive exotic felids (Felis spp.): A multiple regression analysis. *Zoo Biology* **10**, 95–110 (1991).
23. Wielebnowski, N. C., Fletchall, N., Carlstead, K., Busso, J. M. & Brown, J. L. Noninvasive assessment of adrenal activity associated with husbandry and behavioral factors in the North American clouded leopard population. *Zoo Biology* **21**, 77–98 (2002).
24. Burgdorf, J. & Panksepp, J. Tickling induces reward in adolescent rats. *Physiology and Behavior* **72**, 167–173 (2001).
25. Davis, H. & Pérusse, R. Human-based social interaction can reward a rat's behavior. *Animal Learning & Behavior* **16**, 89–92 (1988).
26. Mellor, D. J. Animal emotions, behaviour and the promotion of positive welfare states. *New Zealand Veterinary Journal* **60**, 1–8 (2012).
27. Taylor, K. D. & Mills, D. S. The effect of the kennel environment on canine welfare: A critical review of experimental studies. *Animal Welfare* **16**, 435–447 (2007).
28. Gourkow, N., Hamon, S. C. & Phillips, C. J. Effect of gentle stroking and vocalization on behaviour, mucosal immunity and upper respiratory disease in anxious shelter cats. *Preventive Veterinary Medicine* **117**, 266–275 (2014).
29. Beck, A. M. The biology of the human–animal bond. *Animal Frontiers* **4**, 32–36 (2014).
30. Kahn, P. H. Developmental psychology and the biophilia hypothesis: Children's affiliation with nature. *Developmental Review* **17**, 1–61 (1997).
31. Wilson, E. O. *Biophilia* (Cambridge, MA: Harvard University Press, 1984).
32. Kruger, K. A. & Serpell, J. A. Animal-assisted interventions in mental health: Definitions and theoretical foundations. In Fine, A. H. (Ed.), *Handbook on Animal-Assisted Therapy: Theoretical Foundations and Guidelines for Practice* (Cambridge, MA: Academic Press, 2006), pp. 21–38.
33. Mormann, F. et al. A category-specific response to animals in the right human amygdala. *Nature Neuroscience* **14**, 1247–1249 (2011).
34. Wells, D. L. The effect of videotapes of animals on cardiovascular responses to stress. *Stress and Health* **21**, 209–213 (2005).
35. Edwards, N. E. & Beck, A. M. Animal-assisted therapy and nutrition in Alzheimer's disease. *Western Journal of Nursing Research* **24**, 697–712 (2002).
36. Ainsworth, M. S. Infant–mother attachment. *American Psychologist* **34**, 932 (1979).
37. Bowlby, J. *A Secure Base: Clinical Applications of Attachment Theory.* Vol. 393 (Oxfordshire, UK: Taylor & Francis, 2005).
38. Crawford, E. K., Worsham, N. L. & Swinehart, E. R. Benefits derived from companion animals, and the use of the term "attachment". *Anthrozoös* **19**, 98–112 (2006).
39. Udell, M. A. R. & Brubaker, L. Are dogs social generalists? Canine social cognition, attachment, and the dog-human bond. *Current Directions in Psychological Science* **25**, 327–333 (2016).
40. Vitale, K. R., Behnke, A. C. & Udell, M. A. R. Attachment bonds between domestic cats and humans. *Current Biology* **29**, R864–R865 (2019).
41. Solomon, J., Beetz, A., Schöberl, I., Gee, N. & Kotrschal, K. Attachment security in companion dogs: Adaptation of Ainsworth's strange situation and classification procedures to dogs and their human caregivers. *Attachment & Human Development* **21**, 389–417 (2019).
42. Barker, S. B. & Wolen, A. R. The benefits of human–companion animal interaction: A review. *Journal of Veterinary Medical Education* **35**(4), 487–495 (2011).
43. Callaghan, P. & Morrissey, J. Social support and health: A review. *Journal of Advanced Nursing* **18**, 203–210 (1993).
44. Flannery Jr, R. B. Social support and psychological trauma: A methodological review. *Journal of Traumatic Stress* **3**, 593–611 (1990).
45. Rohlf, V. & Bennett, P. Perpetration-induced traumatic stress in persons who euthanize nonhuman animals in surgeries, animal shelters, and laboratories. *Society & Animals* **13**, 201–219 (2005).

46. Scotney, R. L., McLaughlin, D. & Keates, H. L. A systematic review of the effects of euthanasia and occupational stress in personnel working with animals in animal shelters, veterinary clinics, and biomedical research facilities. *Journal of the American Veterinary Medical Association* **247**, 1121–1130 (2015).

47. Arluke, A. Managing emotions in an animal shelter. In Manning, A. & Serpell, J. (eds.), *Animals and Human Society* (London, UK: Routledge, 1994).

48. Reeve, C. L., Rogelberg, S. G., Spitzmüller, C. & Digiacomo, N. The caring-killing paradox: Euthanasia-related strain among animal-shelter workers. *Journal of Applied Social Psychology* **35**, 119–143 (2005).

49. LaFollette, M. R., Riley, M. C., Cloutier, S., Brady, C. & Gaskill, B. N. Laboratory animal welfare meets human welfare: A cross-sectional study of professional quality of life, including compassion fatigue, in laboratory animal personnel. *Frontiers in Veterinary Science* **7**, 114 (2020).

50. Davies, K. & Lewis, D. Can caring for laboratory animals be classified as Emotional Labour? *Animal Technology and Welfare* **9**, 1–6 (2010).

51. Stamm, B. H. *The Concise ProQOL Manual* (Pocotello, ID: ProQOL.org, 2010).

52. Herzog, H. Ethical aspects of relationships between humans and research animals. *ILAR J* **43**, 27–32 (2002).

53. Graham, M. L. *Working on the 3 Rs: Utilization of Refinement to Enhance the Value of Translational Research in Nonhuman Primates* (Utrecht, The Netherlands: Utrecht University, 2011).

54. Davis, H. Preparing for pleasure, preparing for pain: Animals using human beings as predictors. In de Jonge, F. H. & van den Bos, R. (Eds.), *The Human-Animal Relationship: Forever and a Day* (Assen, the Netherlands: Van Gorcum, 2005), pp. 82–97.

55. Wolfle, T. Laboratory animal technicians: Their Role in Stress Reduction and Human-Companion Animal Bonding. *Veterinary Clinics of North America: Small Animal Practice* **15**, 449–454 (1985).

56. Cloutier, S., Panksepp, J. & Newberry, R. C. Playful handling by caretakers reduces fear of humans in the laboratory rat. *Applied Animal Behaviour Science* **140**, 161–171 (2012).

57. Bassett, L. & Buchanan-Smith, H. M. Effects of predictability on the welfare of captive animals. *Applied Animal Behaviour Science* **102**, 223–245 (2007).

58. Weiss, J. M. Somatic effects of predictable and unpredictable shock. *Psychosomatic Medicine* **32**, 397–408 (1970).

59. Gouveia, K. & Hurst, J. L. Reducing Mouse Anxiety during Handling: Effect of Experience with Handling Tunnels. *PLoS ONE* **8**, e66401 (2013).

60. Hurst, J. L. & West, R. S. Taming anxiety in laboratory mice. *Nature Methods* **7**, 825-U1516 (2010).

61. Hawkins, P. et al. *Refining Rabbit Care: A Resource for those Working with Rabbits in Research* (Hertfordshire: RSPCA, West Sussex and UFAW, 2008).

62. Marr, J., Gnam, E., Calhoun, J. & Calhoun, J. A non-stressful alternative to gastric gavage for oral administration of antibiotics in rabbits. *Lab Animal* **22**, 47–49 (1993).

63. Prescott, M. J. et al. Refining dog husbandry and care. *Laboratory Animals* **38**, 1–94 (2004).

64. Tallet, C., Brajon, S., Devillers, N. & Lensink, J. 13 - Pig–human interactions: Creating a positive perception of humans to ensure pig welfare. In Špinka, M. (Ed.), *Advances in Pig Welfare* (Duxford, UK: Woodhead Publishing, 2018), pp. 381–398.

65. Prescott, M. J. & Buchanan-Smith, H. M. Training nonhuman primates using positive reinforcement techniques. *Journal of Applied Animal Welfare Science* **6**(3), 157–161 (2003).

66. Hemsworth, P. H. & Coleman, G. J. Changing stockperson attitudes and behaviour. In Hemsworth, P. H. & Coleman, G. J. (eds.) *Human-Livestock Interactions, 2nd Edition: The Stockperson and the Productivity and Welfare of Intensively Farmed Animals* (Wallingford, UK: CABI, 2011), pp. 135–152.

67. LaFollette, M. R., Cloutier, S., Brady, C., Gaskill, B. N. & O'Haire, M. E. Laboratory animal welfare and human attitudes: A cross-sectional survey on heterospecific play or "rat tickling". *PLoS ONE* **14**, e0220580 (2019).

68. Burgdorf, J., Moskal, J. R., Brudzynski, S. M. & Panksepp, J. Rats selectively bred for low levels of play-induced 50 kHz vocalizations as a model for Autism Spectrum Disorders: A role for NMDA receptors. *Behavioural Brain Research* **251**, 18–24 (2013).

69. Hall, L. E., Robinson, S. & Buchanan-Smith, H. M. Refining dosing by oral gavage in the dog: A protocol to harmonise welfare. *Journal of Pharmacological and Toxicological Methods* **72**, 35–46 (2015).

2 A Culture of Care

Thomas Bertelsen and Penny Hawkins

CONTENTS

BACKGROUND: WHY A "CULTURE OF CARE?"

The term "Culture of Care" is widely used in a number of different contexts, including hospital trusts,[1] schools[2] and, increasingly, within animal research and testing.[3,4] There is no agreed definition of the Culture of Care in the context of research animal care and use, but a working concept has been set out, including elements relating to appropriate attitudes, behaviors, mindsets, and mutual respect between staff with different roles.[4]

Whatever the setting, the term should be used thoughtfully; for example, one user-establishment's website says, "we are committed to ensuring an excellent Culture of Care," without defining what this actually means. Furthermore, few establishments currently assess their Culture of Care or its impact on the 3Rs (Reduction, Refinement, and Replacement of animal use), animal welfare, scientific quality, or staff morale.[5] It is essential to be open regarding how far your institution has progressed with defining, implementing, assessing, and communicating about the Culture of Care. Otherwise, there is a risk that the concept can become a meaningless "buzzword" (as in "highest welfare standards"), which would be regrettable, since a good Culture of Care should benefit animal welfare, staff morale, science, and corporate accountability. Assessing the Culture of Care, e.g., using a survey-based approach, will help to ensure that it has an ongoing, positive impact on animal and human welfare, science, and corporate social responsibility.

In the context of this book, the primary reason for working with the Culture of Care is an ethical one; it is the right thing to do. There should also be benefits for the 3Rs, scientific translatability and validity, and members of staff at the user-establishment. This includes not only those working in the animal house, e.g., scientists, animal technologists, care staff, and veterinarians, but also members of ethics or animal care and use committees, and staff who may not directly be involved with animal use. It is also essential that top management visibly endorses and encourages the Culture of Care, including developing and maintaining supportive structures and sound leadership.

Recognizing the connection between human and animal well-being, and the need to optimize both, corporate social responsibility (CSR) policies and statements relating to sustainability and workforce well-being are increasingly including elements of animal welfare and the organization's Culture of Care. However, the United Nations Industrial Development Organization does not currently list animal welfare as a key CSR issue,[6] which is an omission in the authors' view, especially given the increasing recognition that the welfare of sentient animals should be given due consideration in a range of sectors.[7] There is also evidence that significant proportions of the population globally are concerned about the use and welfare of animals in research and testing. Opinion polls repeatedly demonstrate that there are expectations that animals will be used only when "necessary," that no avoidable suffering will be caused, and that every effort will be made to find, and use, humane alternatives.[8]

Some of the key views and expectations expressed by the "public" map on to laws and regulations that control animal use, such as European Union (EU) Directive 2010/63/EU[9] and the US Animal Welfare Act.[10] However, the Culture of Care is essentially about going beyond the letter (or minimum standards) of the law and capturing the spirit of the legislation. In the case of the EU Directive, aspects of this are spelled out in the Recitals, which set the context for the Directive. The Recitals acknowledge that animals have intrinsic worth, that the public has ethical concerns about animal use, and that animals should always be treated as sentient creatures.

Local committees, such as animal welfare bodies and ethics committees, play a pivotal role in driving and developing the Culture of Care and moving it on to embrace a "Culture of Challenge." The animal welfare body (AWB) should "foster a climate of care," as developed in a working document from the European Commission on AWBs and national committees and described as a Culture of Care.[3] Bodies that perform similar tasks to the AWB, such as the US Institutional Animal Care and Use Committee (IACUC) and local ethics committees, should also be able to develop the local culture.

This chapter aims to provide inspiration, and encourage reflection, on the institution's current and desired future Culture of Care. Frequent use is made of terms from the European Union (EU) legislative framework and the US Guide for the Care and Use of Laboratory Animals, which emphasizes the importance of "a culture of care focusing on the animals' well-being." However, the principles and approaches set out within this chapter apply globally and should – if not already in place – be adopted by all those involved with the care and use of animals in research and testing.

WHAT IS THE CULTURE OF CARE?

There is no universal definition of a Culture of Care, as every establishment has its own, individual culture. However, a working concept has been set out (Box 2.1), on the basis of the relevant European Commission (EC) working document[3] and guiding principles from a UK multi-stakeholder working group.[11] Establishments need to interpret and implement the factors listed below within their own organizations, with a clear vision of what a Culture of Care means for them. The Culture of Care should not be viewed as a goal in and of itself, but as a means to apply the factors listed in Box 2.1. A strong Culture of Care also has an intrinsic value because it supports both job satisfaction and self-esteem.[5]

The authors of this chapter advocate for a culture that demonstrates caring and respectful attitudes and behaviors towards animals and encourages acceptance of responsibility and accountability in all aspects of animal care and use. This requires more than just responding to externally imposed standards; both management[12] and "frontline" staff should be actively committed to improving the 3Rs, animal welfare, and research, and should work together to achieve this.[13]

**BOX 2.1   SOME KEY FACTORS WHICH BLEND
TOGETHER TO FOSTER A CULTURE OF CARE**

* Appropriate behavior and attitude towards animal research from all key personnel.
* A corporate expectation of high standards with respect to the legal, welfare, 3Rs, and ethical aspects of the use of animals, operated and endorsed at all levels throughout the establishment.
* Shared responsibility (without loss of individual responsibility) towards animal care, welfare, and use. Those with specified roles know their responsibility and tasks.
* A proactive (rather than reactive) approach towards improving standards.
* Effective communication throughout the establishment on animal welfare, care, and use issues.
* The importance of compliance is understood.
* Care staff and veterinarians are respected and listened to and their roles and work are supported throughout the establishment.
* All voices and concerns at all levels throughout the organization are heard and dealt with positively.

The Culture of Care can thus be considered in terms of *what we do and what we think*, i.e., our behavior, our mindset, and how these both shape the establishment's inherent values and attitudes towards the animals it uses and who are in its care.

There are parallels with the Culture of Care concept within the human healthcare sector, addressing how healthcare professionals care for people in vulnerable states and provide for their physical needs. In this sense, "care" means to take *care of*. Another meaning of "care" is to *care about*, e.g., in the sense that "I care about animal welfare when I design my animal studies." Both meanings bear relevance for the care and use of laboratory animals. Animal technologists and care staff, scientists, veterinarians, and animal welfare officers should care for animals by providing food and water, health, welfare, and protection from avoidable harm and suffering (within the potential constraints of the scientific objectives). This should all obviously be in compliance with relevant legislation, but if the Culture of Care is good, people will also seek to go beyond the minimum requirements, adopting a more challenging and animal-centric approach.

To give the culture a strategic direction, it is essential that top management is seen to endorse and encourage the culture, which includes developing supportive structures and fostering corporate and/or institutional legitimacy.[14] Leadership therefore needs to speak of responsibility, empathy, and compassion, which are qualities that are not explicitly addressed in laws, regulations, and standard operating procedures (SOPs). A balance between doing things right (e.g., an SOP) and doing the right things (e.g., performance standards that focus on outcomes) requires continuous leadership attention and alignment within the organization.

Employees should be receptive to these values, which can be facilitated by consulting staff, empowering them, and educating them as to how they can put the company's values into practice. Consequently, empowerment must be delegated to appropriate levels within the user-establishment to employees who will accept this charge within a defined framework of resources and mandate.

Results in terms of implementing the 3Rs, good animal welfare, and robust ethical review can naturally be achieved without a strong Culture of Care, e.g., by a managerial drive. However, the downside of such a managerial drive is that these outcomes may not become embedded in the organization and so will not become part of the culture.

IMPLEMENTING A CULTURE OF CARE

Changing an existing culture requires the willingness to invest in observing and analyzing the status quo, defining a desirable, future state, and projecting a realistic road map of how to get there. The following sections set out some different perspectives to help persuade colleagues to buy in to a Culture of Care.

THE PEOPLE PERSPECTIVE

The primary objective of the Culture of Care is reducing animal use and suffering and improving welfare. However, the aspect of caring for staff working with laboratory animals is also relevant and important. Fortunately, caring for animals and caring for people go hand-in-hand very well when the culture effectively embraces both aspects.[15]

It is a fair assumption that staff working hands-on with animals, e.g., animal care staff, animal technologists, and veterinarians, actually like animals and care about them. Performing procedures that cause discomfort or harm to the animals, and ultimately humanely killing them,[16] is emotionally stressful and puts these people in a difficult situation from an ethical standpoint.

It is important to retain and support caring staff, so a Culture of Care that boosts staff morale and well-being is also more likely to succeed in animal welfare achievements. Some key elements towards this are listed and explained below; see also below where we discuss surrogate markers.

Integrity	To be able to work and act with integrity means that you are able, in practice, to follow the moral principles laid out within e.g., the 3Rs, EU, and national legislation, and user-establishment statements. You can "walk-the-talk" and your colleagues do the same.
Meaning	To understand the purpose of the procedures and the studies in which you participate using animals is fundamental. Understanding and accepting the potential benefit of the research you do, or support (e.g., developing drugs for people with debilitating diseases, assessing the safety of substances, adding to human knowledge), is rewarding and helps to deal with emotional labor.
Influence	The ability to impact and improve any procedure or housing condition for the animals feels like a great relief and it gives staff a perception of making a difference.
Social support	To raise an animal welfare concern, or to bring a novel 3Rs initiative to the table, requires a safe environment and colleagues who listen to you and support you. This is especially the case when the issue raised will require changes to established routines and practices. Social support is an important factor in the development of coping mechanisms.[17]
Recognition	To be recognized for a job well done is something that is widely appreciated. Recognitions can include positive and constructive feedback, financial rewards, and events in the user-establishment where special initiatives are awarded. Recognitions can have their origin from peers and from management, both having important impact if given with consideration.

THE 3Rs PERSPECTIVE

A good Culture of Care includes clear corporate expectations relating to the 3Rs, which ought naturally to lead to their more effective implementation, which is essential from an animal-centric perspective. Although it is nice for humans to work in a culture in which they feel valued and respected, we can assume that an individual laboratory animal would be more concerned about experiencing less suffering (Refinement), having better welfare (Refinement), and ideally not being used in scientific study at all (Replacement). Progress with implementing all 3Rs is a critical outcome measure for a Culture of Care, which should be factored into in-house assessments of the development of the Culture.

An element of Refinement in particular is the "animal-centric" perspective. In a Culture of Care this includes, but is not limited to, the capacity for empathy, appropriate (critical) anthropomorphic thinking and willingness to challenge established methods, addressing qualities like responsibility, empathy, and compassion. These are qualities that are not explicitly addressed in laws, regulations, and SOPs, but which come to the fore in the nature and character of bonds between humans and laboratory animals.[18]

A caring culture towards laboratory animals requires the capacity for empathy. The days of Descartes' thinking, that animals were merely machines and the sound of a whimpering dog in pain was equal to the sound of a creaking door, are fortunately long gone. However, you may still (albeit rarely) observe remnants of this antiquated thinking, where some believe that pain or fear are not present, because they cannot observe visible signs. The increasing literature on animal "pain faces" has demonstrated that indicators may be present, but not previously recognized, which has helped to dispel these attitudes and beliefs.[19–21] The application of human behavior change theory can be helpful in achieving a more progressive, animal-centric Culture of Care; the organization Human Behavior Change for animals has a web page listing helpful resources.[22]

When there is scientific evidence of the presence of nociceptors, complex neural tissues, endogenous opiates, and behaviors that indicate sentience, the approach should be one of informed anthropomorphism in the sense that "if this will inflict pain or harm in humans, it will also do so in the animal" unless there is evidence to the contrary.

THE GOOD SCIENCE PERSPECTIVE

As for the 3Rs, it is also to be expected that a good Culture of Care will lead to good science. It has long been recognized that better welfare equals better science – because animals who are experiencing avoidable negative feelings, such as pain, discomfort, anxiety, fear, stress, or distress, will be mounting physiological responses in an attempt to cope, which consequently will introduce confounds.[23–25]

The scientific upside to a good Culture of Care is that animals with less pain, fear, or discomfort will deliver more statistically robust data and repeatable and reproducible results. Recently, questions about reliability, repeatability, and reproducibility have received considerable attention from the scientific community.[26] An example of this is the use of pre-emptive and appropriate analgesia in surgical procedures; if analgesia is not appropriately provided this will result in a state of surgical stress, where a variety of cytokines are released which are likely to act as confounds.

The scientific narrow vision on so-called well-established animal models may lead to "Cinderella thinking." In the original fairytale by the Grimm brothers the two stepsisters had to chop off their toes and heels in order to fit the glass shoe. Applying this analogy, it is tempting for a scientist to use an animal for a certain "'model," even though this leads to a deficit in terms of animal welfare. The reversed approach would be to modify and adapt (refine) the model to fit the animal such that the negative impact on welfare is avoided or minimized as much as possible.

It is widely acknowledged that many laboratory animals instinctively conceal indicators of pain or distress, for example because they are "prey." Likewise, dogs build bonds with their "owners"/ the caretakers and will go to great lengths not to disappoint them, which may lead to a situation where signs of a non-cooperative behavior are missed. Factors such as these and the absence of empathy may lead to sub-optimal use of analgesics, or otherwise insufficient attention to animals' needs.[27]

A culture that encourages a more animal-centric approach to housing, husbandry, and procedures is thus likely to reduce the prevalence and impact of experimental confounds, leading to more valid and translatable science. In principle, this should be an outcome measure for the Culture of Care, although we acknowledge that quantifying this would be both complex and a long-term goal (see *Assessing the Impact of the Culture of Care*).

METHODS AND TOOLS TO SHAPE A CULTURE

Revisiting the notion that culture comprises "what we think and what we do" clearly emphasizes the "people" aspect. Nevertheless, organizational supporting structures are imperative in order to ensure that the culture has the right "direction" and that employees subscribe to a shared set of values. There are many different "tools" to help achieve this (Box 2.2).

BOX 2.2   TOOLS TO HELP DEVELOP AND SHAPE A CULTURE OF CARE

- Focus groups on specific areas (e.g., refinement), including a wide composition of competencies.
- A scientific review board with the remit of assessing the validity and applicability of the animal models, e.g., by using systematic reviews and meta-analyses.
- An internal 3Rs award to appreciate employees who strive to bring the user-establishment's commitment to the 3Rs into action.
- A Culture of Care award for individuals or initiatives that have successfully challenged and changed aspects of their local, institutional culture to develop or support care and welfare.
- 3R newsletters, highlighting initiatives that benefit and promote animal welfare as well as scientific achievements.
- Supporting (and even rewarding) good role models – employees who demonstrate the animal caring aspect in both actions and voice.
- Encouraging and fostering better communication between scientists and animal technologists/care staff ("Communication and a Culture of Care").

As mentioned above, visible endorsement and accountability from management are required to give initiatives like those in Box 2.2 legitimacy and encouragement. The top-level policies and statements give direction, level of ambition, and empowerment to act on all levels within an aligned framework of resources and mandates.

THE ROLE OF LOCAL ETHICS OR ANIMAL CARE AND USE COMMITTEES

Local committees, such as AWBs, Animal Welfare & Ethical Review Bodies (AWERBs), IACUCs, and ethics committees play a pivotal role in driving, and developing, the Culture of Care. EC guidance on the AWB states that it should work with senior management to ensure that there are structures in place to promote the Culture of Care, keeping these under review to make sure that outcomes are delivered effectively.[3]

The AWB and other committees with similar remits ought to be in a good position to foster effective two-way communications and openness throughout the establishment, which are essential for a good Culture of Care. With this in mind, it is helpful to include the Culture of Care within the terms of reference of your AWB/local committee, e.g., a task of the UK AWERB is to "help to promote a Culture of Care within the establishment and, as appropriate, the wider community."[11] Communications, and inclusiveness, are promoted by incorporating members with a wide range of experience, expertise, and perspectives, at different levels of seniority, and from different parts of the organization. Independent, or lay, members can ask especially insightful questions, which can be very helpful in stimulating the committee to consider proposed projects, and animal housing, from an animal-centric perspective.[28]

A good Culture of Care will be promoted most effectively if efforts are made to engage scientists with their local committee and its processes, members, and activities. For example, if your committee reviews project applications, and the scientists' only contact is to attend while their projects are being reviewed, relations may be less than optimal. Encouraging scientists to present their work to the committee (and/or other staff) outside of project review, to attend meetings, to invite the local committee to visit the study, and to participate in initiatives such as 3Rs awards, or activities to define the local Culture of Care, will all help to achieve and maintain good relations. For committees that include ethical review, providing a "forum for ethical discussion" can help staff to explore their viewpoints and feelings about animal use, improving both engagement and openness.[29]

CULTURE OF CHALLENGE

In the wake of the concept of the Culture of Care set out in EU Directive 2010/63/EU in 2010, the idea of a Culture of Challenge soon followed.

With the professional approach to building the right Culture of Care getting a stronger footing across Europe in its animal facilities, it is time to reach to the next level. Moving from good to brilliant science, and taking Refinement to its true potential, it is time to roll out a Culture of Challenge.[30]

This statement reflects the fact that attitudes towards animals are not static; biological science, animal welfare knowledge, and societal concerns are constantly evolving.[31] Consequently, the view that "we have always done things this way" must repeatedly be challenged in order to achieve continuous improvements within housing, husbandry, *in vivo* procedures, analgesia, etc. A Culture of Challenge is a tool that must be deployed but used with consideration and reason as repeated changes of anything and everything will result in frustration, confusion, and potentially loss of a shared identity, potentially leading to opposing sub-cultures.

The challenge of the status quo must therefore take into account: (i) animal welfare impact; (ii) scientific impact; and (iii) feasibility – how easy will it be to implement the change? Is there a reasonable likelihood of success? Does it require special resources?

Some changes will be in the category of "just do it," whereas others will require carefully planned and orchestrated actions. Common for both is the appropriate involvement of all relevant stakeholders and ensuring that changes are communicated in an efficient and timely manner.

ASSESSING THE CULTURE OF CARE

Establishing a starting point, or the current state of your Culture of Care, is essential if you want to monitor the effects of actions to improve the current culture. Although the previously mentioned EC working document does not explain how "culture" can be assessed, some user-establishments and NGOs have issued questionnaires and surveys in an attempt to gather insight in this matter[5] (Culture of Care network members, personal communication). There are many ways to perform such assessments, and it is important to select an approach that will work effectively for your establishment.[32]

One of the authors has experience with using a comprehensive and multi-tiered approach to assessing his user-establishment's Culture of Care.[33] The term "assessing" is used deliberately, because "analyzing" alludes to an approach based on an established and scientifically approved methodology. The approach outlined below should provide a set of data that will enable the identification of areas of development, and ways of introducing the right activities to guide and direct the culture towards the intended future state.

It is profoundly inspired by the principles of "Social Capital," which Robert D. Putnam has described in his book *"Bowling Alone."*[34] It describes the relations in social networks and where these are characterized by reciprocity, trust, and cooperation. The "outcomes" of the network are not for the individual members, but for a common good.

The translation of this into an animal research setting would be:

- A description of the caring intentions of a user-establishment's employees and how the different internal relations unite and characterize the culture.

- Converting the combination of these two elements into a caring approach primarily for the benefit of the laboratory animals, but also to the members themselves, as Culture of Care is about caring for staff too.

In short, the "capital" in this context is a caring culture which benefits and progresses animal welfare.[35]

WHAT MATTERS

The perception of what a "good" Culture of Care looks like is likely to have as many manifestations as the number of respondents you are asking. It is therefore necessary to describe the Culture of Care using tangible terms that are commonly understood. If we subscribe to the definition of culture as "what we think and do," and recognize that culture has a strong element of relations, we can look at relevant values and relations. Based on a slightly modified version of the concept of Social Capital, the following three value-based notions can be used: Collaboration, Trust, and Integrity.[36] The following six operational topics can also be deployed: Influence, Meaning, Predictability, Social support, Rewards/recognition, and Resources. All of these we will call "surrogate markers." Their specific meaning in relation to Culture of Care will be detailed in the *Surrogate Markers* section. The reader will probably recognize most of the surrogate markers as keywords in many articles and surveys on the Culture of Care. These surrogate markers help to describe the actual culture in more tangible terms, but the different relations that are present within a culture are also important.

There will be many relations within a culture, but we will focus on three of these: bonding, bridging and linking (Figure 2.1). It is important to realize that these relations are different in their nature and their strength. Within the group the relations are usually quite informal and rarely "stated," but quite strong. Between the groups the relations often become more formalized, i.e., agreements are made on how to cooperate.[37] The relations to management are usually quite formalized with established procedures and structures for communication, and perceived "breaches of trust" take a lot of effort to remedy. This is where "walk the talk" is key.

- *Bonding* is the relations within a group of e.g., animal technologists/care staff that unites them.
- *Bridging* is the relation between two or more individual groups, e.g., a group of animal technologists and a group of scientists.
- *Linking* is the relation between one or more individual groups and the management level of the organization.

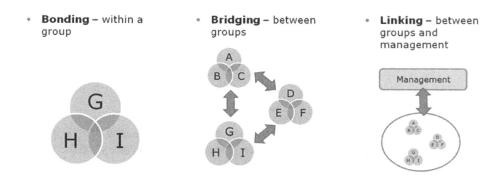

FIGURE 2.1   The main three types of different relations within a culture. It is important to acknowledge these differences, e.g., when addressing potential communication issues. Moving from bonding through bridging to linking the relation becomes gradually more formalized.

The suggested survey approach consequently operates from the "IGLO" model to explore the different relations, as it addresses the Individual, the Group, the Leadership, and the Organization. Another dimension within this is the formation of sub-cultures within a user-establishment. The most evident sub-culture differences are between the different job functions, e.g., animal technologists, veterinarians, scientists, etc. and consequently the survey should also identify the respondent's main job function.

There is a fourth and essential relation in this setting – the human-animal relationship or bond. This has extensively been addressed as early as 2002 where the ILAR Journal published several articles on this topic, some referenced in this chapter. This is essential to enable qualities like responsibility, empathy, and compassion to be exercised.[38,39] This relation unfortunately may have a serious downside which is compassion fatigue,[40] i.e., indifference to the state and well-being of the animals, which may be a result of numerous – but unsuccessful – attempts to improve or change things. A Culture of Care with strong elements of social support and recognition can often reduce or even eliminate this state.[41]

SURROGATE MARKERS

The three value-based notions listed in the previous section have the following meaning in an animal welfare context:

- Collaboration: you are dependent on your colleagues, other groups, your manager, etc. It is a collaborative effort.
- Trust: it is essential that concerns can be voiced without fear of retribution.
- Integrity: this is essentially the user-establishment's "walk the talk" – "we do what we say we will do."

The six operational topics have the following meaning:

- Influence: is it possible for the individual or a group of employees actually to change things and challenge the status quo? The key point here is – are they empowered? Naturally, this empowerment will have to be within an aligned framework of resources and mandates.
- Meaning: is it apparent for the employee(s) that what they do makes sense on a higher level, e.g., "my efforts to improve animal welfare will help develop new medicines/generate new knowledge/assess the safety of new substances (for example) and thus help other people."
- Predictability: to know what will happen if an employee brings forward a new animal welfare initiative, or raises an animal welfare concern, gives a sense of reassurance and control.
- Social support: it is important to know if your new animal welfare initiative will be met with an indifferent shrug or with a listening and appreciative attitude.
- Rewards/recognition: are your animal welfare efforts noticed? And do you get positive feedback? Or are there other kinds of rewards?
- Resources: are there resources available in terms of person-hours and/or in terms of a budget for implementing ideas?

The surrogate markers are very useful when you – after the survey – have follow-up dialogues with the different stakeholders, as it helps you to focus on the essence or nature of a potential issue – to identify what the problem is about in a broader sense.

BUILDING A SURVEY

The following is one example to provide inspiration regarding how to build a survey for your own user-establishment. The multi-tiered approach will enable you to perform an assessment, which is not scientifically validated, but which will be sufficiently accurate to allow you to draw some sound and relevant conclusions.

So – you have the three levels: 1) the different groups of employees, 2) the surrogate markers, 3) the IGLO/different relations.

- First, you ask the participants to indicate which main category of employee they belong to, e.g., animal technologist, lab technician, veterinarian, scientist, license holder, manager, AWB, or ethics committee member, person responsible for ensuring compliance with the legislation.
- Address only one of the surrogate markers per question.
- Group a section of questions relating to the "I" – to the individual. End the section with a field for free-text comments. Do likewise for the "G" – the group you are working with/belonging to, the "L" – the leadership (allow for discrimination between immediate manager and the level(s) above) and for the "'O" – the organization. Decide if all surrogate markers are relevant for each section.

The recommendation is to present the questions like this – "to which extent do you agree with the following statement," and have a choice of ticking "strongly agree," "agree," "neutral," "disagree," and "strongly disagree." To illustrate this multi-tiered approach a few hypothetical examples are presented and explained below.

Example 1: "I contribute to the welfare of the laboratory animals."
In this case, the surrogate marker is "Influence." It is an "I" level in the IGLO.
The result could look like the example shown in Figure 2.2.

Example 2: "In our group we can discuss animal welfare issues with confidence."
The surrogate marker in this case is "Trust." It is a "G" level in the IGLO as it addresses the group.

Example 3: "Management ensures that roles and responsibilities in terms of animal welfare are clear and exact."
The surrogate marker is "Predictability." It is an "L" level in the IGLO as this is about leadership.

Each user-establishment will have its own and unique characteristics, so no single survey or questionnaire will be suitable for all. However, the proposed approach contains some features that have some universal value and should be considered when designing each individual survey.

Great care should be given to the qualitative replies (from the text/comment sections of the survey) from staff with jobs involving hands-on work with animals as they frequently have great insight and an innovative approach especially to refinement techniques.[42]

ASSESSING THE IMPACT OF THE CULTURE OF CARE

As mentioned earlier, an effective and proactive Culture of Care should lead to both direct and indirect benefits for animal welfare, the 3Rs, and the quality of the science at the establishment. There are a number of documents that include potentially useful indicators relating to animal welfare, the 3Rs, and staff relations aspects of the Culture of Care, including the EC working documents on Animal Welfare Bodies and National Committees[3] and Inspections and Enforcement[43], plus the Royal Society for the Prevention of Cruelty to Animals/Laboratory Animal Science Association (RSPCA/LASA) Guiding principles on good practice for AWERBs[11] and the UK regulator's advisory note on identification and management of patterns of low-level concerns at licensed

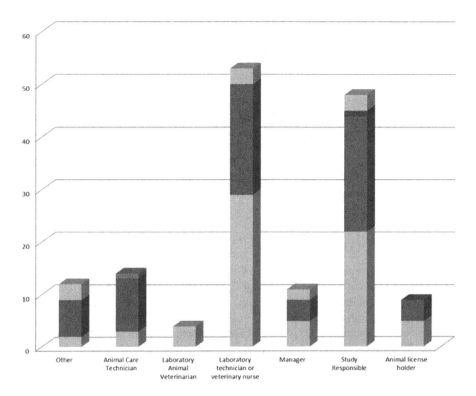

FIGURE 2.2   The number of responses from six different job functions (the category "other" includes cleaning staff and administrative personnel). The graph reflects the differences within a given job function, thus indicating different sub-cultures. The general picture shows that the vast majority agree or strongly agree with the statement "I contribute to the welfare of the laboratory animals." The question addresses the "I" level, the individual, (as opposed e.g., to the group level) and it addresses the surrogate marker "Influence."

establishments.[44] Examples of parameters from these documents which may be helpful in assessing the Culture of Care are listed in Box 2.3.

BOX 2.3   EXAMPLES OF INDICATORS THAT THE CULTURE OF CARE IS LEADING TO POSITIVE OUTCOMES AND IMPACTS FOR ANIMALS AND STAFF

- A sound, consistent environmental enrichment strategy.
- Good staff knowledge and awareness of the 3Rs.
- Ongoing education and training in animal care and welfare, accessible to and encouraged for all levels of staff.
- Engagement with the animal welfare science community.
- Attending veterinarian visits regularly and proactively provides advice and training relating to aseptic surgery, anesthesia, and analgesia.
- Clear audit trails of communications between scientists and animal technologists.
- A clear system is advertised whereby concerns can be raised, including anonymously.
- Good planning and experimental design including advice from a biostatistician to minimize wastage/good communications with users.

Indicators of scientific benefits are less straightforward, but efforts should still be made to consider what these might be for a particular establishment or laboratory, and how these might be assessed – acknowledging that this may be a long-term endeavor. If we assume that there will be fewer experimental confounds, leading to more valid and translatable science, then this could be evidenced by more positive comments from peer reviewers, a higher publication and funding rate, greater efficiency with respect to progressing potential new drugs to clinical trials, greater accuracy with safety assessments, and so on. The AWB or ethics committee could take a lead and convene a workshop, or think tank, to consider what would be feasible and how this could be followed up.

Note that the authors are not suggesting *all* of the above questions are used at every establishment, as this would obviously lead to an overly complex and time-consuming assessment process. We propose a thoughtful combination of a qualitative survey, to obtain people's views, with some more quantitative outcome/impact measures, all within a realistic and manageable framework that does not exceed the available resources.

PROMOTING THE CULTURE OF CARE

Providing staff with information about the local Culture of Care, and how it is developing, is obviously key to ensuring that the culture is meaningful, impactful, and challenging. This can be easily achieved by adding relevant information to current media such as newsletters and the organizational intranet, including email addresses (or physical post boxes). All staff should be encouraged to give feedback, and ask questions, about the Culture of Care. Information can also be included in communications on the AWB, 3Rs, welfare improvements, policy changes, and roles of animal technologists, care staff, training persons, and veterinarians, as recommended in the EC Working Document on AWBs and national committees.[3] As a bottom line, all staff should know that the establishment has set out a local Culture of Care, and where they can find out about this.

NATIONAL MEETINGS ARRANGED BY AUTHORITIES

Within the EU, national committees for the protection of animals used for scientific purposes must be in place in each member state. National committees advise the AWBs on matters dealing with the care and use of animals, ensuring that good practice is shared. Regular meetings arranged by the national committees, as occurs in some member states, are ideal for raising awareness of the importance of user-establishment's work with the Culture of Care. They can also serve as a vital forum for exchanging ideas, raising challenges, and for sharing successes. Outside the EU, the same tasks could be assumed by similar governmental bodies or organizations such as Public Responsibility in Medicine and Research in the US.

SCIENTIFIC CONGRESSES AND MEETINGS

Although meetings held by laboratory animal science associations are often open to include a range of issues, it can be difficult to get animal welfare, ethical, or Culture of Care-related topics on the agenda within "mainstream" scientific congresses. But it is important to keep suggesting talks or workshops (or even mini-symposia) on the Culture of Care, otherwise there will be no progress with respect to raising awareness and support for the concept and all the benefits it can bring. Even if you are unsuccessful in getting the topic on the agenda of a meeting, presentations and posters can mention the local Culture of Care and how this influences your own work.

ORGANIZATIONS AND NETWORKS

Some countries have set up networks for local ethics committees, AWBs, and similar bodies. For example, the Netherlands, Belgium, Denmark, and Portugal have all set up AWB networks, and the

UK national committee has established a regional network of AWERB hubs, which communicate with the AWERBs in their region and the national committee. Networks like these can include initiatives relating to the Culture of Care when ethics committees and AWBs share good practice and compare ways of working.

An international network specifically on the Culture of Care was also launched at the 2016 Federation for Laboratory Animal Science Associations (FELASA) Congress in Brussels. The intention of the network is to provide a forum for the quick and efficient dissemination of ideas and efforts to create a Culture of Care, and to promote a mindset and behavior that continuously and proactively works to advance laboratory animal welfare and the 3Rs. The network has a working concept that includes aiming for more than a culture of compliance and encouraging a Culture of Challenge, rather than accepting established practice. More can be found on the Norecopa website.[4]

OTHER OPPORTUNITIES TO PROMOTE THE CULTURE OF CARE

Many countries now have a movement to increase openness about animal use, for example the UK Concordat on Openness on Animal Research and transparency agreements in Spain and Portugal. The UK Concordat requires organizations to be open about the benefits, harms, and limitations of their animal use in public-facing statements. Statements like these are obvious opportunities to promote the Culture of Care, provided that they are honest (e.g., not saying the Culture is "outstanding" when the facility only complies with minimum legal requirements) and include specific information, beyond simply stating that "there is a Culture of Care."

CONCLUSION

This chapter aims to stimulate reflection on the Culture of Care, including recognizing and promoting its value, observing, and assessing it within an institution, and helping to promote the concept more widely. The authors hope that the reader will feel inspired to start a conversation about progressing the Culture of Care within their institution, for example via the AWB, IACUC, ethics committee, or AWERB, and explore whether the above approaches to shaping and assessing culture could be used to maintain or improve practice. Endorsement from management will be critical, and it may be helpful to use networks such as those listed in *Organisation and Networks* to discuss ways of persuading colleagues and building support if this is not initially forthcoming. The networks can also be used to share information about approaches to assessing the Culture of Care, and we hope that this chapter will be a catalyst for productive liaisons and initiatives that will benefit humans and other animals alike.

REFERENCES

1. Rafferty, A. M., Philippou, J., Fitzpatrick, J. M. & Ball, J. *Culture of Care Barometer. Report to NHS England on the Development and Validation of an Instrument to Measure "Culture of Care" in NHS Trusts* (London, UK: National Nursing Research Unit, 2015). Available at: https://www.england.nhs.uk/wp-content/uploads/2015/03/culture-care-barometer.pdf. (Accessed 14th August 2019)
2. Cavanagh, T., Macfarlane, A., Glynn, T. & Macfarlane, S. Creating peaceful and effective schools through a culture of care. *Discourse: Studies in the Cultural Politics of Education* 33(3), 443–55 (2012). doi: 10.1080/01596306.2012.681902
3. *A Working Document on Animal Welfare Bodies and National Committees to Fulfil the Requirements under the Directive* (European Commission, 2014). Available at: http://ec.europa.eu/environment/chemicals/lab_animals/pdf/endorsed_awb-nc.pdf (Accessed 14th August 2019)
4. Culture of Care. *Norecopa.* Available at: https://norecopa.no/more-resources/culture-of-care (n.d.) (Accessed February 21, 2019)
5. Robinson, S. et al. The European Federation of the Pharmaceutical Industry and Associations Research and Animal Welfare Group reflections on 'Culture of Care' in the context of using animals for scientific purposes. *Laboratory Animals*, 1–12 (2019). doi: 10.1177/0023677219887998

6. What Is CSR. UNIDO (n.d.). Available at: https://www.unido.org/our-focus/advancing-economic-comp etitiveness/competitive-trade-capacities-and-corporate-responsibility/corporate-social-responsibility -market-integration/what-csr (Accessed January 15, 2019)

7. Mills, G. Government must act now on sentience. *The Veterinary Record* **182**(17), 472 (2018).

8. Clemence, M. & Leaman, J. *Public Attitudes to Animal Research in 2016* (London, UK: Ipsos MORI, 2016). Available at: https://www.ipsos.com/sites/default/files/2016-09/Public_attitudes_to_animal_res earch-2016.pdf (Accessed 14th August 2019)

9. Directive 2010/63/EU of the European Parliament and of the Council of 22 September 2010 on the Protection of Animals Used for Scientific Purposes. *Official Journal of the European Union* **276**, 33–79 (2010).

10. Office of the Federal Register (US). *Code of Federal Regulations Title 9 Animals and Animal Products: Parts 1 to 199, Revised January 1, 2011* (Government Printing Office, 2011). Available at: https://ecfr.io /Title-09/cfrv1#0 (Accessed 14th August 2019)

11. Jennings, M. et al. *Guiding Principles on Good Practice for Animal Welfare and Ethical Review Bodies. A Report by the RSPCA Research Animals Department and LASA Education, Training and Ethics Section* (3rd edition) (RSPCA/LASA, 2015).

12. Weichbrod, R. H. et al. *Management of Animal Care and Use Programs in Research, Education, and Testing* (2nd edition). (Boca Raton, FL: CRC Press/Taylor & Francis, 2018), pp. 11–24.

13. Brønstad, A. & Berg, A. G. T. The role of organizational culture in compliance with the principles of the 3Rs. *Lab Animal* **40**(1), 22–26 (2011).

14. Kane-Urrabaza, C. Management's role in shaping organizational culture. *Journal of Nursing Management* **14**, 188–194 (2006).

15. Chang, T & Hart, L. Human-animal bonds in the laboratory: How animal behavior affects the perspective of caregivers. *ILAR Journal* **43**(1), 10–18 (2002).

16. Emotional well-being of staff involved in the killing of animals. 2nd RSPCA/LASA/LAVA/IAT AWERB-UK meeting - 15 June 2017. Available at: https://www.rspca.org.uk/webContent/staticImages/ AWERBUK2017/IAT_workshop.pdf (Accessed 14th August 2019)

17. Narver, H. L. et al. Tributes for animals and the dedicated people entrusted with their care: A practical how-to guide. *Lab Animal* **46**, 369–372 (2017).

18. Russow, L.-M. Ethical implications of the human-animal bond in the laboratory. *ILAR Journal* **43**(1), 33–38 (2002).

19. Häger, C. et al. The sheep grimace scale as an indicator of post-operative distress and pain in laboratory sheep. *PLoS ONE* **12**(4), e0175839 (2017). doi: 10.1371/journal.pone.0175839

20. Reijgwart, M. L. et al. The composition and initial evaluation of a grimace scale in ferrets after surgical implantation of a telemetry probe. *PLoS ONE* **12**(11), e0187986 (2017).

21. Miller, A. L. & Leach, M. C. 2015. The mouse grimace scale: A clinically useful tool? *PLoS ONE* **10**(9), e0136000 (2015). doi: 10.1371/journal.pone.0136000

22. Human Behaviour Change for Animal Welfare. *Human Behaviour Change for Animals* (n.d.). Available at: http://www.hbcforanimals.com/hbc-aw-resources.html (Accessed February 25, 2019)

23. Carbone, L. Pain in laboratory animals: The ethical and regulatory imperatives. *PLoS ONE* **6**(9) (2011), e21578. doi: 10.1371/journal.pone.0021578

24. Gaskill, B. N. & Garner, J. P. Stressed out: Providing laboratory animals with behavioral control to reduce the physiological effects of stress. *Lab Animal* **46**, 142–145 (2017).

25. Gouveia, K. & Hurst, J. Optimising reliability of mouse performance in behavioural testing: The major role of non-aversive handling. *Scientific Reports* **7**, 44999 (2017).

26. Davies, G. F. et al. *Review of Harm-Benefit Analysis in the Use of Animals in Research* (2017). Available at: https://ore.exeter.ac.uk/repository/bitstream/handle/10871/31153/Review%20of%20harm%20benefit %20analysis%20in%20use%20of%20animals.pdf?sequence=4&isAllowed=y (Accessed 14th August 2019)

27. Herzog, H. Ethical aspects of relationships between humans and research animals. *ILAR Journal* **43**(1), 27–32 (2002).

28. Jennings, M. & Smith, J. *A Resource Book for Lay Members of Ethical Review and Similar Bodies Worldwide* (3rd edition). (RSPCA, 2015)

29. Hawkins, P. & Hobson-West, P. *Delivering Effective Ethical Review: The AWERB as a "Forum for Discussion."* (RSPCA, 2017)

30. Louhimies, S. Refinement Facilitated by the Culture of Care. In *Proceedings of the EUSAAT Congres*, Linz, Austria. (2015).

31. Klein, H. J. & Bayne, K. A. Establishing a culture of care, conscience, and responsibility: Addressing the improvement of scientific discovery and animal welfare through science-based performance standards. *ILAR Journal* **48**(1), 3–11 (2007).

32. Hawkins, P. & Bertelsen, T. 3Rs-related and objective indicators to help assess the culture of care. *Animals* **9**(11), 969 (2019). doi: 10.3390/ani9110969

33. Bertelsen, T. Measuring Culture of Care – a practical example. In EUSAAT Conference *2018, Linz, Austria*. Abstract available at: https://eusaat-congress.eu/ (Accessed 14th August 2019)

34. Putnam, R. D. *Bowling Alone: The Collapse and Revival of American Community* (New York, NY: Simon & Schuster Ltd., 2001).

35. Social capital. *Wikipedia.* Available at: https://en.wikipedia.org/wiki/Social_capital (Accessed 14th August 2019).

36. Hasle, P. et al. Organisational social capital and the relations with quality of work and health – a new issue for research. In *ISOCA 2007.* International Congress on Social Capital and Networks of Trust on *18–20, October 2007*, Jyväskylä, Finland.

37. Davies, G. F., Greenhough, B. J., Hobson-West, P., Kirk, R. G. W., Applebee, K., Bellingan, L. C. Developing a collaborative agenda for humanities and social scientific research on laboratory animal science and welfare. *PLoS ONE* **11**(7), e0158791. doi: 10.1371/journal.pone.0158791

38. Wolfe, T. L. Introduction. *ILAR Journal* **43**(1), 1–3 (2002).

39. Bayne, K. Development of the human-research animal bond and its impact on animal well-being. *ILAR Journal* **43**(1), 3–9 (2002).

40. Compassion Fatigue: The Cost of Caring. Human emotions in the care of laboratory animals. *AALAS.* Available at: https://www.aalas.org/media/ca75e87c-8310-4475-bf7e-2bb21d6306e1/naf1YA/EDU/Cost ofCaring.pdf (Accessed 14th August 2019)

41. Shyan-Norwalt, M. R. The human–animal bond with laboratory animals. *Lab Animal* **38**(4), 132–136 (2009).

42. Greenhough, B. & Roe, E. Exploring the role of animal technologists in implementing the 3Rs: An ethnographic investigation of the UK University Sector. *Science, Technology, & Human Values* **43**(4), 694–722 (2018).

43. A working document on Inspections and Enforcement to fulfil the requirements under the Directive. European Commission. (2014) Available at: http://ec.europa.eu/environment/chemicals/lab_animals/p df/endorsed_inspection-enforcement.pdf (Accessed January 15, 2019)

44. Identification and management of patterns of low-level concerns at licensed-establishments. *ASRU.* (2015). Available at: https://assets.publishing.service.gov.uk/government/uploads/system/uploads/attac hment_data/file/512098/Patterns_low-level_concerns.pdf (Accessed 14th August 2019)

3 Animal Emotions and Moods
Why They Matter

Karolina Westlund and Sylvie Cloutier

CONTENTS

INTRODUCTION

Emotions are adaptive. They evolved to enable animals to evaluate and differentiate between different stimuli and respond with specific response patterns adapted to the particular species and environmental conditions. Thus, emotions help animals survive, reproduce, and thrive. Why, then, are emotions interesting to people interacting with laboratory animals, such as biomedical scientists, veterinarians, and animal caretakers?

Emotional experiences are important because they govern decision making, impact personality development and physiology. By taking animals' emotional experiences into consideration, we can improve their welfare in captivity, and get thriving, happy animals that trust people. What might be less obvious is that we can also improve the quality of scientific data by integrating the animal's emotional experiences into the planning of care, husbandry, and research. For example, petting or stroking rats should be avoided because it has been found to induce production of 22 kHz vocalizations associated with negative affective states, especially when not habituated to handling.[1] However, using techniques known to induce positive affective states such as rat tickling could facilitate and accelerate habituation to handling and procedures. Tickling rats immediately before an intraperitoneal injection reduces their fear of handling and humans.[2,3]

Ensuring animals have healthy, positive emotional experiences is not about coddling, it is about ensuring good animal welfare and safeguarding the quality of scientific studies.

PANKSEPP'S SEVEN CORE EMOTIONS

The Core Emotions concept was identified and described by Jaak Panksepp and coworkers over decades of experimental studies involving humans, other mammals, and birds. In a nutshell, using electrostimulation, psychostimulating drugs, and later positron emission tomography (PET) scan among others, Panksepp and his colleagues compared the responses to various stimuli in homologous parts of the brain of different species. They were able to trigger seven distinct emotional reactions – each emotional circuitry found in a specific location within the most ancient, central, parts of the brain, the sub-cortical regions of the mammalian brain, also called the limbic system. This system includes the hypothalamus, the parahippocampal cortex, the amygdala, and several interconnected areas (septum, basal ganglia, nucleus accumbens, anterior insular cortex, and retrosplenial cortex).[4] Panksepp termed these seven behavioral clusters Core Emotions, given that he could only trigger them by stimulating Core parts of the brain. The location of these Core emotional systems in the brain suggests that they evolved a long time ago and that at a basic emotional and motivational level, all mammals are more similar than they are different. These similarities thus include their capability to experience emotions.[4] It can also be inferred that animals with similar brain, physiological, and neuronal systems, e.g., other vertebrates such as birds and fish, experience similar basic emotions. It may be worth noting that as early as 1871, Charles Darwin stated:

> The fact that the lower animals are excited by the same emotions as ourselves is so well established, that it will not be necessary to weary the reader by many details.[5]

The core part of the brain is also where other feelings such as homeostatic affects (bodily states such as hunger and thirst) and sensory affects (i.e., the pleasures and pains of externally provoked sensation such as sweetness and bitterness, disgust, and certain types of pain) are generated.[6] However, in this chapter we focus on Panksepp's seven Core emotional systems.

The seven Core Emotional systems are: CARE, GRIEF, PLAY, LUST, SEEKING, FEAR, and RAGE. They were named using capital letters to clearly distinguish them from the vernacular terms.[6] Panksepp suggested that these seven are a conservative estimate of the actual number of Core Emotions; there may actually be more – but they have at present not been identified, or agreed upon by the scientific community (but see Panksepp[6]). These Core Emotions make sense from the evolutionary perspective – each has tremendous adaptive significance. Also, each Core Emotion has, in diverse scientific studies, been shown to have far-reaching effects on physiology, behavior and welfare.[4,6]

CARE

The CARE system is epitomized by maternal devotion.[4] Even if the nurturing urge is strongest in females in some species, the urge to nurture emanates from inherent brain circuitry shared by both sexes.[4] Blood transfusion from a nursing rat mother and electrical stimulation of the CARE circuitry of the brains of animals were shown to instigate caregiving, nurturing behavior.[4] These would look different depending on the different species, and CARE circuitry would typically be the most vigorously induced in females (especially in postpartum compared to virgin ones) than males. Observable behaviors resulting from stimulation of the CARE system might involve behaviors such as nest building, recovering young, grooming, and licking, as well as lactation and allowing young to suckle. In other words, the beneficiary of the CARE system is primarily the offspring, not the individual experiencing the emotional state.

Studies have demonstrated the importance of good maternal CARE. The quality of mothering is one of the factors that most influence an animal's personality development, through epigenetic effects. For instance, the degree to which a rat mother licks and grooms her offspring (a function of her own stress level) influences how stress-sensitive and explorative those offspring become as adults.[7,8]

In summary, the CARE system ensures that mothers, in the case of most laboratory species, can properly rear their offspring. If a mother is stressed, her caretaking abilities will be impaired, and her infants risk becoming easily stressed as they grow up. If she is calm, she can properly care for them, and they will grow to become more exploratory and confident – and less susceptible to stress.

Preventing stress is crucial in a laboratory animal environment, since stress is a potential confound in biomedical research. To ensure happy, confident laboratory animals, environments should be set up to optimize maternal care. Additionally, actively selecting breeding females demonstrating good maternal skills might improve the success of breeding programs.

GRIEF

GRIEF (formerly named PANIC) is the source of non-sexual attachment, or stated differently, the separation-distress mechanism of the mammalian brain. It is most often expressed as separation-related vocalizations and arousal in young animals, and depressive-like symptoms in adults.[4] GRIEF has survival value in that when a young animal is separated from the primary caregiver, those vocalizations and agitated arousal attract the mother to her youngsters' location, so she can retrieve them before predators find them.

Social animals strive for social contact. Particularly for young ones, interactions with other individuals may be the difference between life and death. Not only does close physical contact help young animals thermoregulate, it also involves the release of feel-good neurotransmitters and hormones that contribute to solidifying social bonds and are crucial to normal brain development.

In more applied terms, to minimize the negative impacts of the GRIEF system breeding programs should look over weaning practices (e.g., at which age are young separated from their mother or family group), strive to eliminate single housing of social species and encourage housing in appropriately sized groups. For instance, differential development in areas of the frontal cortex between juvenile rats housed in pairs or in groups of four has been reported[9] and it has been demonstrated that some stereotypic behaviors are associated with maternal deprivation.[10] Also, escape-related behavior is more prevalent in rodents separated early as compared to those weaned late.[11]

PLAY

PLAY is the source of joyous interactions. Expression of PLAY is more prevalent in young animals who play more than adults, but PLAY is experienced at all ages. PLAY has tremendously beneficial effects on brain development, personality, social skills, and stress sensitivity. Social play, in particular, helps in calibrating emotional reactions to events, frightening and unexpected situations,[12,13] and induces resilience to depression and anxiety.[14]

PLAY requires the appropriate conditions for full expression. For example, it can be depressed by fear and hunger.[6] Thus, it is a good indicator of welfare because it is sensitive to negative environmental and emotional states (although exceptions were noted, see Ahloy-Dallaire et al.[15]): if young animals do not exhibit play behavior, it is likely that something about the environment is suboptimal. Also, if they do not feel safe, they will not play.

To put this into practice, laboratory environments should encourage PLAY-related behavior in growing animals. This may be accomplished by providing sufficient space and enrichment structures that encourage the expression of locomotor, object, and social play behaviors; providing shelters – making animals feel safe so that play may be expressed. Environments conducive to the expression of play behaviors can help young animals "practice for the unexpected" and cope with stressful situations.[12,13] Additionally, playful human-animal interactions, such as rat tickling, may be beneficial in inducing positive emotional states, especially in relation to humans, boosting confidence and reducing the stressful impact of further handling and biomedical procedures (Figure 3.1).[16,17]

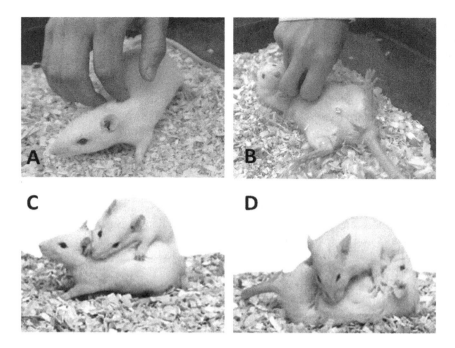

FIGURE 3.1   Using rat tickling humans can mimic aspects of the rats' playful rough-and-tumble behaviors such as dorsal contact (A and C) and pins (B and D) during handling. This playful handling which triggers the PLAY system like rough-and-tumble play, consistently gives rise to positive behavioral responses, and also increases playful behaviors.[13]

LUST

LUST is the system leading the elaboration of sexual desire and attachment in the brain. This is one of the systems where strong sex differences are observed. Sexual urges govern male and female behavior somewhat differently, males typically searching for many mating opportunities – their reproductive success is related to the number of females that they can impregnate. Females, on the other hand, are more interested in finding the healthiest and most attractive mates: they prioritize quality rather than quantity.

An important, albeit indirect, effect of the LUST system on laboratory animals is associated with housing practices. The most obvious one would perhaps be the difficulty in group-housing intact males of some species in the vicinity of stimuli indicating the presence of intact females. For instance, male mice are notoriously difficult to co-house, typically because they tend to show inter-male aggressive behavior from around the age of sexual maturation, which might be exacerbated by the presence of females. This is not surprising, considering that mice do not form multi-male-multi-female societies in the wild, and illustrates the importance of being familiar with the ranging patterns and social structures of the wild ancestors of the species housed in captivity today. Solutions to co-housing males of social species such as mice might be to find low-aggressive strains, ensure high compatibility by keeping siblings together, and reduce the number of incoming stimuli triggering aggression. Moreover, during husbandry procedures the caretaker should ensure the transfer of nesting material during cage changes and avoid the transfer of bedding which includes aggression-inducing pheromones found in urine (Figure 3.2).[18] Alternatively, it is possible to keep them in a social setting consisting of neutered females.[19]

FIGURE 3.2 On cage cleaning day transferring nesting material (from A to D) but not bedding material from the dirty (left) to the clean (right) cage contributes to reducing aggression, especially in male mice. (Photos: Brianna N. Gaskill)

SEEKING

Animals have an urge to explore, investigate, and integrate information from their surroundings with the goal of acquiring all the resources needed for survival, and the Core Emotional SEEKING system is at the heart of these types of behaviors. In the wild, animals typically need to expend a considerate amount of time and energy foraging, which includes searching for, procuring, preparing, and eating food.

Moreover, it has been demonstrated that SEEKING is mainly activated during the searching/handling phase of foraging: the appetitive phase. During the consummatory phase, when animals chew and swallow, SEEKING is all but switched off.

This system is one of the most important emotional systems of the brain because it has connections with many regions of the lower brain and many higher regions of the brain including the frontal cortex. Thus, its activation has a big impact on the brain and the body. Furthermore, it interacts with the other Core Emotional systems. SEEKING is involved in the appetitive phase of all the other emotional systems. Excitement and anticipatory behaviors accompanying the search or expectation for something (which can be food, a mate, one's mother, a territory trespasser, the safety of a nest) are generated by SEEKING.[4]

In captivity, the foraging opportunities available to animals are typically limited, and this is particularly true in the laboratory environment. When food is provided in hoppers, the only thing animals need to do is approach and eat. They do not need to search for food, they do not need to cache it or look for their cache (for hoarding species). They are thus deprived of the appetitive phase of foraging, only experiencing the consummatory mode, meaning that their SEEKING system is typically not activated as much in captivity as in the wild. Thus, for many species, encouraging species-typical foraging behavior has several and varied beneficial effects: increased diversity of behavior, reduced abnormal behavior, and reduced aggression being common effects.[20,21] Developing and using environmental enrichment devices that encourage species-typical foraging behavior – without increased aggression due to limited access to, or monopolization of, resources – is therefore important.

Keeping the SEEKING system active on a regular basis could minimize the development of depressive states as an under-activated system leads to behavioral signs of depression such as lethargy. That is why, in addition to environmental enrichment, training and cognitive tasks, which activate SEEKING are activities that could be added to an enrichment program.

FEAR

Fearful responses help animals survive potentially life-threatening situations, and when the FEAR circuitry is engaged, animals will move slowly, freeze, become immobile, or flee.

Animals often respond fearfully to stimuli that we humans do not even notice. For example, rats are reported to have a hereditary predisposition to exhibit behavioral and physiological stress responses to the odor of certain natural predators such as cats,[22] although habituation has been reported following repeated exposure.[23] Thus, the mere presence of a few cat hairs on an experimenter or caretaker's clothing can induce a strong fear response. Similarly, mice might respond fearfully to the smell of rats – a natural predator,[24] but also habituate as shown by growth and reproduction measures.[25] Understanding triggers of species-specific *innate* fear is important but even more so is preventing the acquisition of *fear learning* through classical conditioning, where animals learn to fear previously neutral stimuli that are reliable predictors of innate fear triggers.

There are at least two major issues involving FEAR that might impact experimental results in the laboratory animal facility. One involves home-cage perceived safety, and the other suboptimal human-animal interactions. The latter risks inducing or increasing fear learning with regards to handling and biomedical or experimental procedures.

In practical terms, unless animals have been acclimatized to handling, they will likely respond with a FEAR reaction when being picked up and handled. It is therefore tremendously important to teach animals not to fear humans, to prevent fear learning and the risk of sensitization, which might impact the validity of scientific data, as well as being an animal welfare concern. Proper acclimatization, involving systematic desensitization and counter conditioning are ways of achieving this (see chapters on animal learning and animal training in this book). Ensuring that the very first interactions between an animal and its caretaker are positive, non-aversive experiences can greatly contribute to reducing the risk of future fear learning (Figure 3.3).

RAGE

The RAGE response centers around anti-predatory violent responses, serving to incapacitate the predator, as well as frustration-induced behavior and aggressive responses related to competition for resources such as food treats, nest site, mate, or triggered by pain related to health issues or experimental conditions.

When untrained animals are caught and restrained, their first reaction is often to kick, scratch, or bite their way out of their predicament. This RAGE response is risky in terms of injuries, and also entails a physiological stress response that may compromise scientific data.

In the laboratory environment, evidence of the RAGE response might be, for example, when mice lifted by their tails turn around to bite their handler. Recent data suggest that mice do not bite when lifted in a cupped hand or receptacle, such as a tunnel.[26]

The use of systematic training (operant and classical conditioning procedures) for health and behavior assessment of animals, can minimize the expression of RAGE-related aggressive behaviors, particularly after surgical procedures.

FIGURE 3.3 Handling, especially for small animals like rats, can have an important impact on their subsequent response to handling for husbandry, care, and testing. Rat tickling, a handling technique that mimics aspects of rat rough-and-tumble play, can effectively habituate rats to handling and prepare them for research procedures within a work week.[13] Following as few as one to three 15 second tickling sessions, rats will calmly sit in their handler's hand. (Photo: Henry Moore Jr., Photographer, Washington State University, College of Veterinary Medicine)

SUMMARY OF THE CORE EMOTIONAL SYSTEMS

In short, the seven Core Emotional systems all involve emotional reactions that serve to improve animals' survival. Four of them are pleasant (CARE, PLAY, LUST, SEEKING) and experiencing them is beneficial to the animal's health and welfare – these emotional states are reinforcing, and animals are generally willing to work to experience them. The other three are unpleasant (GRIEF, FEAR, RAGE), and animals will generally work to avoid them. Moreover, extended exposure to the three unpleasant core emotions are associated with reduced overall well-being. A clear understanding of the Core Emotional systems is necessary to understand higher, more complex, emotional levels and their interactions because all these systems link up with various homeostatic and sensory affective mechanisms including higher, more complex ones.[4]

CORE AFFECT SPACE

Panksepp's model of emotions came about from experimental findings; those seven Core Emotions can be triggered in humans and animals alike by stimulating homologous parts of the brain. Another model, the Core Affect Space, characterizes emotions using a dimensional perspective.[27,28] In this model, rather than specifying and naming different emotional states, emotions are

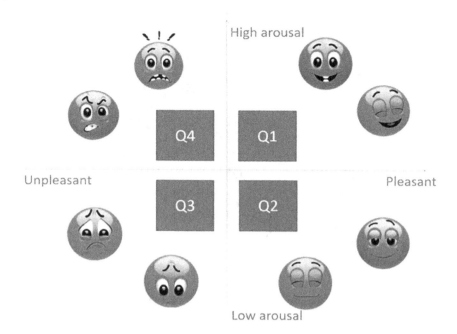

FIGURE 3.4 Graphical view of the Core Affect Space model.

conceptualized in a framework consisting of two continuous and perpendicular dimensions.[29,30] Briefly, *valence* (how agreeable versus disagreeable a specific emotion can be) is depicted on the x-axis, and *arousal* (how reactive the animal experiencing the emotion is to stimuli), on the y-axis (Figure 3.4).

The resulting four quadrants matrix, mapping emotional states, motivations, or sensations, is referred to as the Core Affect Space. Each quadrant includes specific valence and arousal levels. Quadrant 1 (Q1) and Quadrant 2 (Q2) thus both signify pleasant emotions but involving different levels of arousal. In contrast, animals with emotional status located in Quadrant 3 (Q3) and Quadrant 4 (Q4) experience unpleasant emotions. Q3 and Q4 differ in the level of arousal. On the one hand, a predator eagerly sniffing the ground may have a high arousal and experience a pleasant sensation, thus residing in Q1. Once it has eaten, and is comfortably resting, arousal is reduced and so the individual's emotional status may shift to Q2. On the other hand, a prey animal fleeing from a predator might be in Q4, experiencing high arousal and negative emotions. Finally, animals found in Q3 rate low on the arousal dimension and are not enjoying the sensation.

MOODS

Individual emotional responses may be fleeting, spanning mere seconds or minutes. Since an animal is constantly experiencing a trajectory through the Core Affect Space, these moment-to-moment emotional reactions, if they are often repeated, may lead the animal to spend most of its time in a certain area of the Core Affect Space, and experience a specific affective state, or mood. The animal may, for instance, become anxious (spending most of his time in Q4), or depressive (Q3). Moods impact physiology, behavior, and decision making. For instance, anxious and depressed rats tend to make negative judgments about events and to interpret ambiguous stimuli unfavorably.[31] These types of cognitive biases have been used as indicators of affective states in animals and to evaluate welfare.[32]

WELFARE AND POSITIONING IN CORE AFFECT SPACE

It seems reasonable to assume that animals whose moods, or positions in Core Affect Space, reside mainly in Q3 and Q4 have lower welfare than animals experiencing moods predominantly in Q1 and Q2. Prolonged exposure in Q3 could potentially lead to a state of learned helplessness, as the animal learns that important features in the environment are beyond its control.[33] In Q4, an animal experiences emotions characterized by high arousal and negative valence such as frustration, fear, and aggressive episodes which also correlate with reduced welfare (Figure 3.5).[33]

In contrast, Q1 and Q2 are locations in Core Affect Space where animals enjoy positive valence. For example, exploratory behavior, which has been suggested as an indicator of good welfare, can be based in Q1.[34] In Q2, there is an absence of threat, and this quadrant is where recovery may occur.[29] In Q2, the animal does not show frustration, fear, or aggression. Boissy et al.[35] proposed that positive experiences are a core component of good welfare while negative emotions correlate with reduced welfare. Taken together, these findings indicate that Q1 and Q2 are indeed the two quadrants where animals' emotions should predominantly reside to enjoy good welfare, and that prolonged state in emotions associated with Q3 and Q4 are conducive to poor welfare. However, this does not imply that animals should never experience momentary challenges that involve affective loops into Q3-Q4. Indeed, challenging situations may initially have negative valence and be frustrating but might serve to motivate the animal to seek the solution to the current challenge. When the goal is attained, the frustration decreases.[36] In contrast, insolvable or inescapable challenges will prolong the stay in Q3-Q4, and this type of sustained negative emotional state is a potential health and welfare risk.[37]

Positioning in core affect space is not stagnant, but a function of experience.[29,38] Welfare can thus be improved. Humans can provide sensations, motivations and thus, induce discrete emotions contributing to shift an animal's mood states through Core Affect Space to quadrants that are beneficial from a welfare perspective (Q1 and Q2 rather than Q3 and Q4). Shifting an animal to the

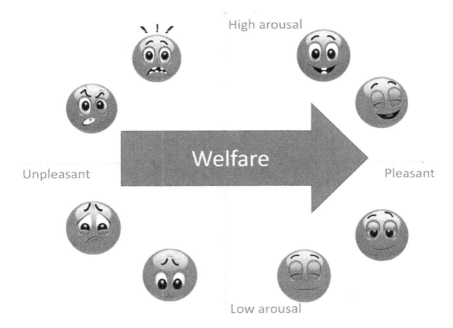

FIGURE 3.5   Positioning welfare in the Core Affect Space model.

right, along the valence axis in the Core Affect matrix, should thus improve welfare, regardless of the level of arousal.

COMBINING THE TWO MODELS OF EMOTION

While Panksepp's CORE Emotion model allowed identification of brain areas and behaviors associated with basic emotions, it did not provide a clear explanation for arousal. For example, piglets utter barks when playing, but also in alarming or startling situations.[39] In both PLAYful and FEARful situations the piglet appears to be in an aroused state. Because variations between juvenile play vocalizations and alarming calls have not yet been identified,[39] the affective state of piglets can only be determined by the context. In such a case, combining the Panksepp CORE Emotion and the CORE Affect Space models can facilitate understanding and interpretation of animal behaviors and emotions. It could also provide alternative explanations for the lack of differences in physiological measurements such as cortisol levels when comparing positive (e.g., playful handling) and negative/neutral (e.g., petting or restraint, and intraperitoneal injection) situations which all involve high arousal (for example see Cloutier et al.[2]). The Panksepp Core Emotion systems can be incorporated into the Core Affect Space matrix as shown in Figure 3.6. When considering the placement of the seven core emotions within Core Affect Space, one might argue that SEEKING, PLAY, and LUST involve medium to high arousal and positive emotional states, which puts them in Q1 of the Core Affect Space matrix. CARE would typically be pleasurable but involve lower arousal, placing it in Q2.

In turn, GRIEF, FEAR, and RAGE are all negative/aversive, and experiencing these emotions would place an animal to the left in Core Affect Space matrix. Although FEAR and RAGE are different systems, they both involve high arousal, hence an animal experiencing any of those emotions would be expected to be in Q4. With regards to GRIEF, the initial separation response typically involves agitation, vocalization, and high arousal (Q4), but with time, if reunion does not occur, the

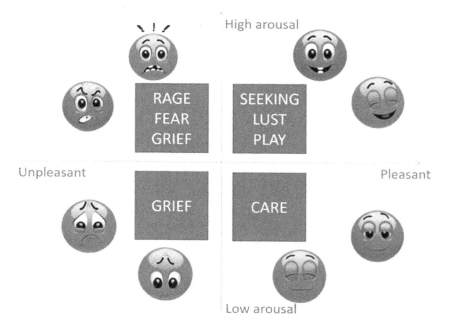

FIGURE 3.6   Graphical view of how Panksepp's Seven Core Emotions and the Core Affect Space model can be combined.

animal may "give up" and become apathetic and withdrawn (Q3). Thus, combining the two models also shows that subjectively different emotions can occupy the same space (even within quadrants) in the Core Affect Space matrix (Figure 3.6).

CONCLUSION

Animals experience emotional states, motivations, sensations, and moods, and are on a constant journey through the Core Affect Space.

Long-term negative emotional states tend to compromise welfare. Thus, we should strive to allow animals to spend most of their time experiencing positive emotions – shifting to the right in the Core Affect Space (Q1 and Q2). To achieve this, we should aim to diminish exposure to GRIEF, FEAR, and RAGE, and encourage SEEKING, PLAY, LUST, and CARE.

Awareness of animals' emotional states and a systematic approach to dealing with them may thus improve both animal welfare as well as the quality of biomedical studies.

REFERENCES

1. Brudzynski, S. M. & Ociepa, D. Ultrasonic vocalization of laboratory rats in response to handling and touch. *Physiology & Behavior* **52**, 655–660 (1992).
2. Cloutier, S., Wahl, K., Baker, C. & Newberry, R. C. The social buffering effect of playful handling on responses to repeated intraperitoneal injections in laboratory rats. *Journal of the American Association for Laboratory Animal Science* **53**(2), 161–166 (2014).
3. Cloutier, S., Wahl, K., Panksepp, J. & Newberry, R. C. Playful handling of laboratory rats is more beneficial when applied before than after routine injections. *Applied Animal Behaviour Science* **164**, 81–90 (2015).
4. Panksepp, J. & Biven, L. *The Archaeology of Mind: Neuroevolutionary Origins of Human Emotions* (New York, NY: WW Norton & Company, 2012)
5. Darwin, C. Comparison of the mental powers of man and the lower animals. In *The Descent of Man and Selection in Relation to Sex*. Cambridge Library Collection – Darwin, Evolution and Genetics, pp. 34–69. Cambridge: Cambridge University Press (2009; first published in 1871).
6. Panksepp, J. *Affective Neuroscience. The Foundations of Human and Animal Emotions* (New York, NY: Oxford University Press, 1998)
7. Meaney, M. J. Maternal care, gene expression, and the transmission of individual differences in stress reactivity across generations. *Annual Review of Neuroscience* **24**, 1161–1192 (2001).
8. Kappeler, L. & Meaney, M. J. Epigenetics and parental effects. *Bioessays* **32**, 818–827 (2010).
9. Bell, H. C., Pellis, S. M. & Kolbet, B. Juvenile peer play experience and the development of the orbitofrontal and medial prefrontal cortices. *Behavioural Brain Research* **207**, 7–13 (2010).
10. Latham, N. R. & Mason, G. J. Maternal deprivation and the development of stereotypic behaviour. *Applied Animal Behaviour Science* **110**, 84–108 (2008).
11. Würbel, H. & Stauffacher, M. Age and weight at weaning affects corticosterone levels and development of stereotypies in ICR mice. *Animal Behaviour* **53**, 891–900 (1997).
12. Pellis, S. & Pellis, V. *The Playful Brain, Venturing to the Limits of Neuroscience* (Oxford, UK: Oneworld Publications, 2009).
13. Špinka, M., Newberry, R. C. & Bekoff, M. Mammalian play: Training for the unexpected. *The Quarterly Review of Biology* **76**, 141–168 (2001).
14. Burgdorf, J., Kroes, R. A., Beinfield, M. C., Panksepp, J. & Moskal, J. R. Uncovering the molecular basis of positive affect using rough-and-tumble play in rats: A role for insulin-like growth factor I. *Neuroscience* **168**, 769–777 (2010).
15. Ahloy-Dallaire, J., Espinosa, J. & Mason, G. Play and optimal welfare: Does play indicate the presence of positive affective states? *Behavioural Processes* **156**, 3–15 (2018).
16. LaFollette, M. R., O'Haire, M. E., Cloutier, S., Blankenberger, W. B. & Gaskill, B. N. Rat tickling: A systematic review of applications, outcomes, and moderators. *PLoS ONE* **12**(4), e0175320 (2017).
17. LaFollette, M. R., O'Haire, M. E., Cloutier, S. & Gaskill, B. N. Practical rat tickling: Determining an efficient and effective dosage of heterospecific play. *Applied Animal Behaviour Science* **208**, 82–91 (2018).

18. Van Loo, P. L. P., Kruitwagen, C. L. J. J., Van Zutphen, L. F. M., Koolhaas, J. M. & Baumans, V. Modulation of aggression in male mice: Influence of cage cleaning regime and scent marks. *Animal Welfare* **9**, 281–295 (2000).

19. Ewaldsson, B. et al. Who is a compatible partner for a male mouse. *Scandinavian Journal of Laboratory Animal Sciences* **42**, 2014–2017 (2016).

20. Bayne, K., Mainzer, H., Dexter, S., Campbell, G., Yamada, F. & Suomi, S. The reduction of abnormal behaviors in individually housed rhesus monkeys (*Macaca mulatta*) with a foraging/grooming board. *American Journal of Primatology* **23**(1), 23–35 (1991).

21. Honess, P. E. & Marin, C. M. Enrichment and aggression in primates. *Neuroscience & Biobehavioral Reviews* **30**(3), 413–436 (2006).

22. Blanchard, R. J., Blanchard, D. C., Rodgers, J. & Weiss, S. M. The characterization and modelling of antipredatory defensive behaviour. *Neuroscience & Biobehavioral Reviews* **14**, 463–472 (1990).

23. Dielenberg, R. A. & McGregor, I. S. Defensive behaviour in rats towards predatory odors: A review. *Neuroscience & Biobehavioral Reviews* **25**, 597–609 (2001).

24. Arndt, S. S., Loharech, D., van't Klooster, J. & Ohl, F. Co-species housing in mice and rats: Effects on physiological and behavioural stress responsivity. *Hormones and Behavior* **57**(4): 342–351 (2010).

25. Pritchett-Corning, K. R., Chang, F. T. & Festing, M. F. W. Breeding and housing laboratory rats and mice in the same room does not affect the growth or reproduction of either species. *Journal of the American Association for Laboratory Animal* **48**(5), 492–498 (2009).

26. Hurst, J. L. & West, R. S. Taming anxiety in laboratory mice. *Nature Methods* **7**, 825–826 (2010).

27. Russell, J. A. Core affect and the psychological construction of emotion. *Psychological Review* **110**(1), 145–172 (2003). doi: 10.1037/0033-295X.110.1.145

28. Russell, J. A. & Barrett, L. F. Core affect, prototypical emotional episodes, and other things called emotion: Dissecting the elephant. *Journal of Personality and Social Psychology* **76**(5), 805–819 (1999).

29. Mendl, M., Burman, O. H. & Paul, E. S. An integrative and functional framework for the study of animal emotion and mood. *Proceedings of the Royal Society B: Biological Sciences* **277**(1696), 2895–2904 (2010).

30. Trimmer, P. C., Paul, E. S., Mendl, M. T., McNamara, J. M. & Houston, A. I. On the evolution and optimality of mood states. *Behavioral Sciences* **3**(3), 501–521 (2013).

31. Harding, E. J., Paul, E. S. & Mendl, M. Animal behaviour: Cognitive bias and affective state. *Nature* **427**(6972), 312 (2004).

32. Bethell, E. J. A "how-to" guide for designing judgment bias studies to assess captive animal welfare. *Journal of Applied Animal Welfare Science* **18**(sup1), S18–S42 (2015).

33. Maier, S. F. Learned helplessness and animal models of depression. *Progress in Neuro-Psychopharmacology and Biological Psychiatry* **8**(3), 435–446 (1984).

34. Basset, L. & Buchanan-Smith, H. M. Effects of predictability on the welfare of captive animals. *Applied Animal Behaviour Science* **102**, 223–245 (2007).

35. Boissy, A. et al. Assessment of positive emotions in animals to improve their welfare. *Physiology & Behavior* **92**(3), 375–397 (2007).

36. Meehan, C. L. & Mench, J. A. The challenge of challenge: Can problem solving opportunities enhance animal welfare? *Applied Animal Behaviour Science* **102**(3), 246–261 (2007).

37. Ursin, H. & Eriksen, H. R. The cognitive activation theory of stress. *Psychoneuroendocrinology* **29**(5), 567–592 (2004).

38. Nettle, D. & Bateson, M. The evolutionary origins of mood and its disorders. *Current Biology* **22**(17), R712–R721 (2012).

39. Chan, W. Y., Cloutier, S. & Newberry, R. C. Barking pigs: Differences in acoustic morphology predict juvenile responses to alarm calls. *Animal Behaviour* **82**, 767–774 (2011).

4 Abnormal Behavior

Andrea Polanco, Jamie Ahloy-Dallaire, and María Díez-León[*]

CONTENTS

A mouse repeatedly executes backflips. A rabbit bites the bars of her cage nonstop. A monkey is known to poke himself in the eye. Those who work with laboratory animals for any length of time will eventually witness these or other disturbing activities belonging to the wide family of "abnormal" behaviors. They will undoubtedly wonder why the animals are acting this way, whether this means there is something wrong with the animals, and how they can make this behavior stop.

We begin this chapter by discussing what abnormal behaviors are and why they should be taken seriously by those in charge of care and husbandry. We then examine the commonalities and differences between forms of abnormal behaviors and between species. Finally, we present some strategies for prevention, management, and treatment of common abnormal behaviors of laboratory animals.

ABNORMALITY

There is no definitive line separating abnormal from normal behaviors, and little consensus on how abnormality itself should be defined.[1] Historically, the term has had both a descriptive function – normal is common or average, abnormal is rare or unusual – and a normative function – normal is healthy or good, abnormal is pathological or wrong.[2]

In this chapter, we largely avoid the descriptive approach. Behaviors are not considered abnormal simply because they are rare or unusual. This mirrors the practice among mental health researchers, who now avoid classifying the unusual as abnormal in part to avoid pathologizing or stigmatizing minority populations.[3] Moreover, abnormal behaviors are in fact very common in some animal populations: for instance, in one study, every single mouse developed repetitive wire-gnawing.[4] By a conservative estimate, 50% of laboratory mice worldwide show some form of abnormal repetitive behavior.[5] See Table 4.1 for definitions of the most common forms of abnormal repetitive behaviors (ARBs) shown in laboratory animals.

Abnormality may also be defined descriptively in terms of difference from a reference population. Behavior of captive animals might be considered abnormal if it deviates qualitatively or quantitatively from that of their wild counterparts.[6,7] However, this definition does not fully capture the meaning of the term as it is used in this chapter either. Animals in captivity face different challenges and opportunities than do those in the wild, so it is not surprising that their time budgets can

---

[*] All authors contributed equally.

TABLE 4.1

Forms of Abnormal Repetitive Behaviors in the Laboratory Animal Species Covered in This Book

Abnormal Repetitive Behavior	Description	Laboratory Animals Affected
Pacing/route-tracing.	In terrestrial animals: walking in the same pattern, back and forth or in a circle. In mice, this can also involve climbing the cage lid. In fish: repetitive swimming, back and forth or in a circle (sometimes called "waving").	Non-human primates.[27,62,90,91,111] Mice.[22,34,46,112,113] Dogs.[117,118] Rabbits.[75] Fish.[114–116]
Spinning/twirling.	Repeated horizontal turns of the body. In mice, this could involve spinning while hanging onto cage lid. In dogs, this may involve tail-chasing.	Non-human primates.[62,90,111,112] Mice.[22,34,46,112,113] Dogs.[117,118]
Somersaulting/ flipping.	Repeated backward flipping of entire body.	Non-human primates.[27,62,90,91,111] Mice.[113,119,120] Rats.[121]
Bouncing/jumping.	Repeatedly jumping up and down.	Non-human primates.[27,62,90,91,111] Mice.[46,113,120]
Head weaving/ swaying.	Repeatedly moving the head side-to-side.	Non-human primates.[27,62,90,111] Rabbits.[122,123] Pigs.[124]*
Rocking.	Repeatedly moving body back and forth or side-to-side while remaining in same spot.	Non-human primates.[27,62,90,91,111]
Swinging.	Repeatedly moving back and forth while holding onto the cage ceiling.	Non-human primates.[27,62,90,91,111]
Saluting/eye-poking.	A "saluting" gesture of the hand over the eye; often involves a digit (frequently the thumb) being pressed into the eye-socket.	Non-human primates.[62,91,111]
Floating limb.	An arm or leg being lifted seemingly without the animal's knowledge; sometimes used to self-groom as though the animal is being groomed by another; often results in self-biting when animal sees limb.	Non-human primates.[111,125,126]
Self-biting/ self-injurious behavior.	Self-biting, scratching or some other form of self-mutilation which may result in injury (although not always).	Non-human primates.[27,62,91,111]
Hair chewing or hair-pulling.	Chewing on his/her own fur (rabbits) or pulling/plucking out one's hair, which may then involve hair ingestion (non-human primates). Called "barbering" in mice and rats (which can involve plucking of the whiskers too).	Non-human primates.[62,91,111] Rabbits.[122] Mice.[43,46,127,128] Rats.[129]
Bar biting/mouthing.	Repetitively biting or chewing on cage bars.	Rabbits.[122,123,130] Pigs.[124]* Mice.[22,34,112] Rats.[129,131,132]
Clawing at cage/ sham digging.	Repetitively scratching or clawing at cage walls or corners.	Rabbits.[122,123,130] Gerbils.[49]
Sham chewing.	Chewing when nothing is present within the mouth.	Pigs.[124]*

*Farmed pigs

For examples in other species see: cats,[77] ungulates.[58] For species for which there is no specific literature available on ARBs, it is reasonable to assume (to an extent, see "Species Differences" section) that they will show a certain degree of similarity to closely related species (e.g., the ARBs of guinea pigs are likely to be similar to those of other rodents). For graphic examples (videos, pictures) of some of the ARBs described here, please visit the following websites: http://animalbiosciences.uoguelph.ca/~gmason/StereotypicAnimalBehaviour/library.shtml (covering several species); http://mousebehavior.org/videos/ (specific to laboratory mice).

be quite different.[8] In addition, while most of the abnormal behaviors discussed in this chapter are rarely if ever seen in the wild, the label "abnormal" is not applied to other novel behaviors seen in captive animals, such as enthusiastically interacting with keepers or artificial items.[9]

Abnormal, as a normative term, is loaded with the assumption that the behavior is somehow problematic or undesirable.[10] In human clinical practice, the framing of mental disorders stresses the presence of dysfunction or of distress in the patient.[1] Similarly, with respect to animals, abnormal behavior can be taken to mean behavior that results from underlying pathology or dysfunction, or somehow reflects distress. For example, stereotypic behavior has conventionally been defined as repetitive, unvarying, and with no apparent goal or function; more recent accounts instead define it in terms of causal mechanisms such as frustration, repeated attempts to cope with a restrictive captive environment, or changes to behavioral control mechanisms in the brain induced by this environment (N.B. stereotypic behavior is also commonly referred to as "stereotypy," although some authors suggest a more restrictive use of this term).[11] Rather than looking at causes of the behavior, some authors focus on its consequences and therefore reserve the terms abnormal or pathological for behaviors that additionally have negative effects on performers or their conspecifics, such as causing distress or physical harm or interfering with normal behaviors such as eating and sleeping.[12] This maps to some extent onto the distinction between malfunctional and maladaptive behaviors, which we return to below. In this chapter, we define abnormal behavior as resulting from underlying distress or pathology (known or at least hypothesized). Some of the behaviors we touch upon do have negative consequences on performers (e.g., self-injurious behavior), but this need not be the case.

The study of abnormal behavior among animals has traditionally centered on ARBs such as stereotypic behaviors, which are performed in consecutive repetitions (e.g., backflipping, body-rocking, or pacing) or on recurrent behaviors with longer, variable intervals of time between repetitions (e.g., excessive grooming or regurgitation and reingestion). More recently, attention has also been paid to forms of excessive inactivity as potential abnormal behaviors associated with depression in captive animals.[13,14] This chapter covers all of these forms of abnormal behavior.

WHY WE SHOULD CARE ABOUT ABNORMAL BEHAVIORS

Abnormal behaviors of laboratory animals are of interest both as a welfare issue and as a scientific matter. Abnormal behaviors in captive animals usually indicate inadequate welfare (understood here primarily as an individual's subjective experience),[15] and performing some of these behaviors can additionally cause or aggravate other health and welfare problems. Preclinical research on mental health uses captivity-induced or experimentally induced abnormal behaviors in animal models to investigate conditions with similar symptoms in humans. Finally, in research where these behaviors are not the topic of study, their presence may impede our ability to use animals as valid models to produce results that are translatable to human patients.

Abnormal behaviors are associated with captive environments that lead to poor animal welfare. In laboratory, farm, and zoo animals, stereotypic behaviors are usually (but not always) more prevalent, or more frequent, in housing and husbandry environments that are worse for welfare,[5] determined using indicators such as fear behavior, physiological signs of stress, or showing avoidance in preference tests. More stereotypic behavior is usually observed in animals housed in barren or "standard" cages than in those housed in quarters provided with environmental enrichment: "an animal husbandry principle that seeks to enhance the quality of captive care by identifying and providing environmental stimuli necessary for optimal psychological and physiological well-being."[16] For example, female vervet monkeys spent more time performing stereotypic behaviors when housed in small, barren laboratory cages than when they were moved to larger cages or provided foraging logs and other enrichments, including access to an exercise cage.[17] Similarly, barbering (see Table 4.1) was more prevalent and more severe in mice living in standard cages compared to enriched cages[18] (N.B. in the neuroscience literature, the term "environmental enrichment" is often used more broadly to refer to any object/addition/modification to housing conditions aimed to increase environmental complexity,[19] irrespective of welfare). Abnormal behaviors can be a sign of current welfare problems, but can also be a persistent symptom of past welfare problems, such as those stemming from inadequate rearing environments. In one extreme example, infant rhesus macaques, who had been individually confined in small vertical chambers with smooth walls for 45 days, showed very low levels of activity and adopted hunched and self-clasping postures for up to nine months after being returned to standard caging.[20] While providing environmental enrichment is typically an effective means of reducing stereotypic behavior, it rarely if ever abolishes it outright,[21] and it can be more effective for younger animals.[22] Indeed, stereotypic behavior, and potentially other abnormal behaviors, can result from captivity-induced changes to brain function which can be difficult to reverse. In fact, they can, in some cases, be thought of as "scars" of past welfare problems.[11] It is also important to recognize that abnormal behaviors are not *automatically* signs of poor welfare. These behaviors may appear in research animals subjected to genetic or pharmacological manipulations that hijack behavioral control mechanisms (e.g., ulcerative dermatitis in SHANK3 mutant mice;[23] stereotypic behaviors in animals given amphetamines[24]). Alternative causes should be ruled out before jumping to the conclusion that apparent abnormal behavior reflects poor welfare: for example, alopecia may result from an autoimmune disease that spontaneously arrests hair follicle growth rather than from hair-pulling,[25] and urine-drinking in diabetic monkeys could simply be a response to glucosuria (sweet-tasting urine).[12]

Some abnormal behaviors can cause or aggravate health and welfare problems. This is clearly the case for various self-injurious behaviors that create lesions or damage other body parts (e.g., eyes) (see Table 4.1). Putting an end to these behaviors should be a priority, even if the chosen solution does not necessarily address the root problem.[26] Humane euthanasia may even be the most appropriate response to rare chronic or recurrent cases of self-injurious behavior that cannot be resolved.[27] Other behaviors have limited or unclear potential to do harm: for example, bar-biting could conceivably result in tooth damage, fur loss from barbering or hair-pulling could impair thermoregulation,[28] and extreme inactivity could worsen obesity-related health outcomes in some animals.[29] Many other behaviors, including most stereotypic behaviors, likely do no physical harm to the animal at all. In fact, a long-standing hypothesis proposes that performing these behaviors helps animals cope with the stress of captivity, though there is no documented evidence of any benefits outside of a few special cases.[30] Within a given housing environment, unlike when comparing environments, the most stereotypic individuals often do have the best welfare according to other indicators,[5] but this is not necessarily so *because* they perform stereotypic behaviors: differences in personality may instead predispose animals to both stereotypic behavior and better welfare in certain environments.[31] Using restraint or further confinement to prevent animals from performing stereotypic behaviors may simply be more stressful than letting them do so,[32] though some experiments indicate no negative or positive effect of preventing stereotypic behaviors.[33] Similarly, absence of these behaviors is not *necessarily* evidence of good welfare, with extreme forms of inactivity appearing as an alternative response to welfare-reducing environmental conditions.[34]

Abnormal behaviors have been studied in animal models of human psychiatric disorders.[35] This includes both captivity-induced behaviors and those induced experimentally by genetic, neurobiological, or pharmacological manipulations. Stereotypic behaviors are present in multiple human disorders.[36] Animal models have been developed to study each of them, with perhaps the most significant focus being on autism (e.g., in rodents[37] and primates[38]). Captivity-induced behaviors that target the integument, such as barbering and ulcerative dermatitis in rodents, have been used as models of trichotillomania and skin-picking disorder respectively.[39,40] Excessive inactivity in primates and mice has been used to model symptoms of depression.[34,41] Abnormal animal behaviors may be superficially similar to many different human disorders, but careful consideration is needed to determine whether they are homologous (retained from a common ancestor), and therefore likely to have shared mechanisms across species. For example, increased self-grooming or barbering has been used to model symptoms of trichotillomania, obsessive-compulsive disorder, autism, Tourette syndrome, schizophrenia, anxiety disorders, and more.[42] However, research has established that barbering in mice most closely matches human trichotillomania in terms of its epidemiology,[43] and that the same treatment is effective in both mice and humans.[39] Abnormal behaviors can be powerful models of human disorders, but only if thoroughly validated.

Finally, it is questionable whether research animals who display abnormal behaviors are appropriate model animals.[44,45] Animals with abnormal behaviors have previously been found to have abnormal physiology[41] and abnormal brain function.[37,46] The main remedy for abnormal behaviors – environmental enrichment – is also known to have positive effects on neurobiology and physiology, including immune function.[42] With that in mind, are abnormally behaving animals, who typically live in non-enriched environments, suitable stand-ins for human patients in biomedical research? Will their bodies and brains react to experimental treatments the same way as those of healthy humans living in rich, complex environments?[47]

MALADAPTIVE OR MALFUNCTIONAL?

If we consider a spectrum of causal factors, at one end we find abnormal behaviors that are performed by animals who are kept in "abnormal" conditions (i.e., environmental conditions they have not evolved in), but who are otherwise – neurologically, physiologically, and anatomically – normal. Their abnormal behaviors result then from attempting to behave in an adaptive, functional way in a non-suitable environment. These are known as maladaptive[48] behaviors. One classic example of maladaptive behavior in laboratory animals is that of gerbils who dig stereotypically in the absence of environmental cues associated to the natural burrowing system that they have evolved to inhabit.[49] This can be considered an example of frustration (at not having a suitable burrow) triggering repetitive behavior. Some forms of aggression in laboratory mice are also an example of maladaptive behavior[50] (see Chapter 8). At the other end of the spectrum we find abnormal behaviors that are performed by animals who regardless, or because of, the environment in which they are kept in are themselves "abnormal." In other words, their behaviors result from dysfunction or pathology, typically at the neural level. These are known as malfunctional[48] behaviors. Examples of malfunctional behaviors in laboratory animals are stereotypic behaviors performed as a result of brain lesions in, for example, rhesus macaques,[51] or the pharmacologically induced repetitive gnawing behavior of rats.[52] Maladaptive behaviors can develop into malfunctional behaviors, but they can equally be eliminated if their root cause is addressed early, before any dysfunction or pathology develops. Maladaptive causal explanations underlie the specific form of abnormal behavior that might be seen (e.g., bar-biting mice, see below), while malfunctional explanations underscore the importance of normal brain development during early life, and thus why treatment of some abnormal behavior (e.g., ARBs) is more effective in younger animals,[22,53] or why individuals who perform ARBs also show cognitive and social impairments.[54–56] In terms of welfare, we might be more concerned about an individual performing ARBs caused by thwarted motivations in their current environment, than an animal showing ARBs due to early life poor welfare conditions when their current environment ensures better welfare (although welfare could be compromised by sheer inability to control one's own behavior[54]).[5]

Therefore, understanding causal mechanisms behind abnormal behaviors, and where in this continuum between malfunctional and maladaptive a given animal is at any given point in time, can help determine both if the animal is a valid research model and whether their welfare is compromised. These two judgments often go hand in hand but may also be dissociated in some cases. As an example of the latter, consider a mouse whose somersaulting is caused by early life conditions that altered neural function, and not by any current captivity-induced frustration. This stereotypic behavior may not be effectively reduced by enrichment, yet this might not *necessarily* raise welfare concerns if the individual does not show any other signs of reduced welfare. However, the mouse's value as a research model may be compromised (e.g., if part of a study assuming healthy, normally developed neurophysiology). In such a scenario, some people may be tempted to stop enriching animals who seem unresponsive; however, the issue with malfunctional behaviors is precisely that they may persist even if the factors that initially elicited them are no longer present. Providing effective enrichment will likely still have positive welfare effects, even if this is not reflected in reduced abnormal behaviors (see above), an issue we return to towards the end of the chapter.

SPECIES DIFFERENCES

While the fundamental principles about abnormal behaviors described above can be generalized across species, we also find several differences between species. These often relate to differences in natural ecology and life history, particularly in terms of differences in the behavior's timing, prevalence, form, efficacy of treatment, and possibly even their neurobiological basis. The following sections are primarily concerned with ARBs, although the principles are likely to be transferrable to other forms of abnormal behavior as well.

Animals are more prone to engage in abnormal behaviors at certain times of day, and the specific time differs between species. Species that would spend a lot of time roaming and searching for food in the wild (e.g., many carnivores) appear to perform these behaviors primarily *before* food delivery, while species that spend a considerable amount of time processing food (e.g., most ungulates, including pigs) are more likely to perform them *after* food is provided.[57,58] Therefore, it is important to be aware of animals' natural foraging behavior when assessing ARB, as the occurrence of ARBs is likely to be affected by timing of food provision.[59] Similarly, it is important to be aware of the circadian activity patterns of a given species in order to be able to identify whether ARBs occur in the first place – observations of laboratory mice in rooms without reverse light cycles are likely to miss any ARB performed at night, when mice are naturally most active.

Related species might be markedly different, even in terms of whether they perform ARBs in the first place. For example, ferrets perform little to none of the ARBs of related species, such as American mink.[60] The degree of domestication might play a role, although we find a similar pattern in two species of rodents – laboratory rats and mice, kept in captivity for similar amounts of time – with rats rarely displaying captivity-induced ARBs, whereas an estimated 50% of laboratory mice on average do show these behaviors.[5] Likewise, apes are more likely to perform ARBs (followed by Old World and New World monkeys, and prosimians),[61] while among Old World monkeys, rhesus, and cynomolgus macaques are more likely to perform motor stereotypic behaviors than baboons.[62]

Species also vary in terms of form (i.e., the specific pattern of behavior) of ARB displayed. Pacing and bar-biting tend to be the most common forms across different laboratory animals, while species-specific forms include self-biting (e.g., non-human primates) and sham digging (e.g., rabbits) (see Table 4.1). Pacing is relatively common in carnivores, and oral stereotypic behaviors in ungulates, reflecting differences in natural foraging styles.[63] As mentioned above, the absence of ARBs does not necessarily indicate neutral or positive welfare states, as some animals may also express poor welfare through excessive inactivity (see Table 4.2).

The neural pathways involved in the performance of certain ARBs might also differ among species. For example the ARBs of deer mice appear to be caused by dysfunction in the dorsal basal

TABLE 4.2
Forms of Excessive Inactive Behaviors in the Laboratory Animal Species Covered in This Book

Excessive Inactivity	Description	Laboratory Animals Affected
Inactive but awake.	Motionless while awake (eyes are open).	Mice.[34] Rats.[133] Ferrets.[102]
Hunched posture.	A slumped body posture, head at or below shoulders, while awake (eyes are open). The animal is not engaged in any activity.	Non-human primates.[13,134,135]

ganglia,[64] while those of C57BL/6 mice appear linked to alterations in the ventral striatum[65] (see e.g., [66-70] for further reading on the neural basis of ARB).

Acknowledging species differences is important for treatment implications as different forms and aetiologies may require different treatments (as reviewed in the last section of this chapter). Similarly, some forms of treatment might be more or less effective for different species. For example, for non-human primates, non-social enrichment during early rearing is not nearly as effective as social enrichment in order to prevent the development of ARBs.[71]

ABNORMAL REPETITIVE BEHAVIOR PREVENTION

To date, prevention (and enrichment) paradigms have not been explored or tested for efficacy to reduce ARBs in fish. As such, this and the following section are only relevant for terrestrial species that exhibit ARBs. Likewise, the treatment of abnormal inactivity has also been researched less extensively than ARBs, so the following sections are mainly focussed on the latter.

Given that early life experiences have a long-lasting influence on the development of ARBs, prevention efforts should focus on the rearing environment. For instance, early removal from the mother is a well-known risk factor in laboratory monkeys,[72,73] rodents,[74] and rabbits,[75] in addition to domestic dogs[76] and cats.[77] As such, decisions regarding when to separate mother and offspring should mimic natural weaning ages appropriate for the species, which is approximately 10–14 months old in monkeys,[78] 20–35 days of age in mice,[79] four weeks of age for rabbits,[80] and 8–16 weeks in cats and dogs.[81,82] Maternal deprivation is the biggest risk factor for ARB development in monkeys: even being reared with peers is not enough to prevent ARBs in nursery-reared monkeys.[83]

However, late weaning is generally not an adopted practice because it can conflict with management objectives, including maximizing breeding output. Thus, extra care should be taken to ensure animals receive a stimulating and rich environment with the primary objective to improve animal welfare. Successful enrichment paradigms applied during the rearing period which have significantly reduced the later development of ARBs in rodents include double-sized cages with additional bedding materials and cylindrical structures such as tunnels and/or cardboard rolls.[84,85] As such, if purchasing animals from vendors, we recommend inquiring about their rearing environment as this could differ between facilities. However, it should be cautioned that removing enrichments from animals reared with these items can sometimes result in more ARBs than in animals who have only been housed in barren standard cages[86] (although not always[87]). Thus, the safest route to guarantee optimal animal welfare would be to ensure that enrichments are permanent once provided, and replaced once worn out or destroyed (i.e., one should avoid removing enrichment once given). Rotating between enrichments on a regular basis (without waiting for them to become worn or dirty) can also be an effective way of preventing habituation and reduced usage over time. Unfortunately, research on enrichment paradigms that successfully prevent the onset of ARBs in other laboratory species (e.g., dogs, pigs, rabbits) is lacking.

But what if we are too late in preventing ARBs? What do we do about the individuals who already are displaying these behaviors?

ABNORMAL REPETITIVE BEHAVIOR TREATMENT

Once an animal has been identified as exhibiting an ARB, it is important to consider the animal's current and prior history of stressors. This is because ARBs can either reflect immediate stressors (e.g., current food restriction triggers oral ARBs in farmed pigs;[88] current single housing triggers ARBs in laboratory primates;[27] current move from group outdoor housing to single indoor housing triggers hunched postures in primates[89]), early aversive experiences (e.g., maternal deprivation as mentioned above triggers ARBs in a variety of species), or cumulative effects over the animal's lifetime (e.g., ARBs in monkeys are predicted by a history of more research projects, blood draws, cage moves, pair separations, and years spent indoor housed and/or single housed[27,90-92]). If the animal does not have a history of aversive experiences or barren housing, then the ARB most likely stems from a current stressor. Well-known examples include bar-chewing in mice (which results from escape attempts and/or a motivation to explore outside the home-cage[93,94]) and stereotypic digging in gerbils (which is caused by the absence of a suitable burrowing system[49]). If the stressor is properly identified, then corresponding enrichments can be given. In the examples above, oral ARBs were the result of food restriction or single housing, which suggests that altering the feeding regime or social housing would be adequate treatments as they would address the respective root causes of the behavior. Likewise, providing an appropriate burrow ceased stereotypic digging in gerbils. In contrast, if the animal has a history of aversive experiences (including long-term barren housing), then more general enrichments can be given to reduce their ARBs.

Environmental enrichment appears in various forms including social, physical, foraging, or a combination. Social housing (physical access to conspecifics) is a basic need that should be met for social species (e.g., monkeys, dogs, pigs, rabbits, and rodents) either through pair or group housing, as echoed by laboratory animal guidelines (e.g., the Guide for the Care and Use of Laboratory Animals[95]) and legislation (e.g., European Directive 2010/63/EU[96]). However, some laboratory protocols may require single housing due to incompatibility issues or research and veterinary protocols. In that case, efforts should be made to ensure that social animals receive positive human interactions with their caretakers, such as positive reinforcement training (animals are rewarded with food for successfully completing a task), which is known to reduce ARBs in monkeys.[97]

Physical enrichment is the provision of play structures and/or manipulable objects including playrooms, perches, hammocks, chew toys, substrates, bedding materials, or a combination of these, that allow animals to explore and engage in more species-typical behaviors, fewer ARBs, and less abnormal inactivity (e.g., nesting materials for rodents;[34,98,99] chewing blocks for rabbits;[100] puzzle feeders for monkeys;[101] tunnels and balls for ferrets[102]). Likewise, foraging enrichment is the provision of species-typical food but given to the animals either scattered in their home-cage or via a type of food puzzle to increase foraging and to reduce ARBs (cats,[103] monkeys,[71] rodents[98]). Overall, it is highly recommended to socially house gregarious laboratory animals, with the addition of consistent physical and foraging enrichments to create a stimulating environment.

For more extreme forms of ARB that involve self-injury, other fast-acting treatments should be considered in addition to environmental enrichment, given the immediate harmful effects of these behaviors. These include pharmaceutical or dietary treatments: for example, antidepressants and antianxiety drugs reduce self-biting in laboratory monkeys (while not affecting non-injurious ARBs,[104,105] although they are known to reduce pacing in zoo-held carnivores,[106] likely by reducing overall levels of activity rather than by addressing the root cause), while antioxidants reduce ulcerative dermatitis.[40] Physical alterations may also effectively prevent harmful consequences of the behavior in specific cases: simply trimming the toenails of mice affected by ulcerative dermatitis reliably resolves dorsal lesions.[26]

Finally, it is important to bear in mind that effectiveness of treatment is likely to depend on the specific causes of the behavior, and thus to recognize that not all ARBs are equally affected by the same type of enrichment (i.e., some enrichments may be more successful in treating some ARBs than others). For example, puzzle feeders and foraging substrates are more effective for suppressing pacing and rocking, but not backflipping, bouncing, and self-directed ARBs in laboratory monkeys.[107,108] In another study, different types of enrichment (positive-reinforcement training, food, non-food, and social) reduced "'part-of-body" ARBs (involving only one part of the body, such as head-nodding), but only social enrichment (i.e., pair housing) and positive-reinforcement training reduced "whole-body" ARBs (e.g., pacing).[109] Likewise, whole-body (e.g., whole-body bobbing and pacing) and head-only ARBs (e.g., head-twirling) are more likely to be reduced by physical enrichments (e.g., plastic toys) than stereotypic scratching on cage walls in mink (a Carnivora model for ARBs).[110] These studies illustrate how some ARB forms are more treatable than others (depending on the enrichment type) in a variety of species. Thus, the form of ARB is important to acknowledge when devising a treatment scheme. Most importantly, if the underlying motivation of the ARB is understood, then tackling that motivation will be the most effective way to reduce and/or prevent it. This might be achieved through careful observation of potential environmental triggers, form, timing, and response to environmental change, as well as the life history of the individual (e.g., early life rearing conditions) performing these behaviors. While causation might be difficult to ascertain for every single individual, it is important to remember that overall, these behaviors are linked to poor welfare, and thus one should err on the side of caution and not ignore them but rather treat them as a warning sign. For this reason, and as mentioned already, it is important to continue to provide enrichment, even if it has no direct observable effects on ARBs or other abnormal behaviors, as enrichment can still be beneficial in other ways.

REFERENCES

1. Miles, S. R. & Averill, L. A. Definitions of abnormality. In Cautin, R. L. & Lilienfeld, S.O. (Eds.), *The Encyclopedia of Clinical Psychology* (Hoboken, NJ: Wiley-Blackwell, 2014). doi: 10.1002/9781118625392.wbecp546
2. Duschinsky, R. & Chachamu, N. Abnormality, overview. In Teo, T. (Ed.), *Encyclopedia of Critical Psychology* (New York, NY: Springer, 2014). https://link.springer.com/referenceworkentry/10.1007/978-1-4614-5583-7_496
3. Morin, S. F. Heterosexual bias in psychological research on lesbianism and male homosexuality. *American Psychologist* 32, 629–637 (1977). doi: 10.1037/0003-066x.32.8.629
4. Würbel, H., Stauffacher, M. & von Holst, D. Stereotypies in laboratory mice – Quantitative and qualitative description of the ontogeny of "wire-gnawing" and "jumping" in Zur:ICR and Zur:ICR nu. *Ethology* 102, 371–385 (1996).
5. Mason, G. J. & Latham, N. R. Can't stop, won't stop: Is stereotypy a reliable animal welfare indicator? *Animal Welfare* 13, S57–S69 (2004).
6. Veasey, J. S., Waran, N. K. & Young, R. J. On comparing the behaviour of zoo housed animals with wild conspecifics as a welfare indicator, using the giraffe (*Giraffa camelopardalis*) as a model. *Animal Welfare* 5, 139–153 (1996).
7. Birkett, L. P. & Newton-Fisher, N. E. How abnormal is the behaviour of captive, zoo-living chimpanzees? *PLoS ONE* 6, e20101 (2011). doi: 10.1371/journal.pone.0020101
8. Fraser, D., Weary, D. M., Pajor, E. A. & Milligan, B. N. A scientific conception of animal welfare that reflects ethical concerns. *Animal Welfare* 6, 187–205 (1997).
9. Markowitz, H. Engineering environments for behavioral opportunities in the zoo. *The Behavior Analyst* 1, 34–47 (1978).
10. Würbel, H. The motivational basis of caged rodents' stereotypies. In Mason, G. & Rushen, J. (Eds.), *Stereotypic Animal Behaviour: Fundamentals and Applications to Welfare* (2nd edition) (Wallingford, UK: CABI, 2006), pp. 86–120.
11. Mason, G. Stereotypies in captive animals: Fundamentals and implications for animal welfare. In G. Mason & J. Rushen *Stereotypic Animal Behaviour: Fundamentals and Applications to Welfare* (2nd edition) (Wallingford, Oxfordshire, UK: CABI, 2006), pp. 325–356.
12. Bayne, K. & Novak, M. Behavioral disorders. In Bennett, B. T., Abee, C. & Henrickson, R. *Nonhuman Primates in Biomedical Research: Biology and Management* (Cambridge, MA: Academic Press, 1998), pp. 485–500.
13. Willard, S. L. & Shively, C. A. Modeling depression in adult female cynomolgus monkeys (*Macaca fascicularis*). *American Journal of Primatology* 74, 528–542 (2012). doi: 10.1002/ajp.21013
14. Fureix, C. & Meagher, R. K. What can inactivity (in its various forms) reveal about affective states in non-human animals? A review. *Applied Animal Behaviour Science* 171, 8–24 (2015). doi: 10.1016/j.applanim.2015.08.036
15. Duncan, I. J. H. Animal welfare defined in terms of feelings. *Acta Agriculturae Scandinavica Section a-Animal Science* Supplement 27, 29–35 (1996).
16. Shepherdson, D. J. Introduction: Tracing the path of environmental enrichment in zoos. In Shepherdson, D. J., Mellen, J. D. & Hutchins, M. (Eds.), *Second Nature: Environmental Enrichment for Captive Animals* (Washington, DC: Smithsonian Institution Press, 1998). pp. 1–12.
17. Seier, J., de Villiers, C., van Heerden, J. & Laubscher, R. The effect of housing and environmental enrichment on stereotyped behavior of adult vervet monkeys (*Chlorocebus aethiops*). *Lab Animal* 40, 218–224 (2011). doi: 10.1038/laban0711-218

18. Bechard, A., Meagher, R. & Mason, G. Environmental enrichment reduces the likelihood of alopecia in adult C57BL/6J mice. *Journal of the American Association for Laboratory Animal Science* **50**, 171–174 (2011).

19. Nithianantharajah, J. & Hannan, A. J. Enriched environments, experience-dependent plasticity and disorders of the nervous system. *Nature Neuroscience* **7**, 697–709. doi: 10.1038/nrn1970 (2006).

20. Suomi, S. J. & Harlow, H. F. Depressive behavior in young monkeys subjected to vertical chamber confinement. *Journal of Comparative and Physiological Psychology* **80**, 11–18 (1972). doi: 10.1037/h0032843

21. Shyne, A. Meta-analytic review of the effects of enrichment on stereotypic behavior in zoo mammals. *Zoo Biology* **25**, 317–337 (2006). doi: 10.1002/zoo.20091

22. Tilly, S. L. C., Dallaire, J. & Mason, G. J. Middle-aged mice with enrichment-resistant stereotypic behaviour show reduced motivation for enrichment. *Animal Behaviour* **80**, 363–373 (2010). doi: 10.1016/j.anbehav.2010.06.008

23. Peça, J. et al. Shank3 mutant mice display autistic-like behaviours and striatal dysfunction. *Nature* **472**, 437–442 (2011). doi: 10.1038/nature09965

24. Randrup, A. & Munkvad, I. Stereotyped activities produced by amphetamine in several animal species and man. *Psychopharmacologia* **11**, 300–310 (1967).

25. Sundberg, J. P., Silva, K. A., Li, R. H., Cox, G. A. & King, L. E. Adult-onset alopecia areata is a complex polygenic trait in the C3H/HeJ mouse model. *Journal of Investigative Dermatology* **123**, 294–297 (2004). doi: 10.1111/j.0022-202X.2004.23222.x

26. Adams, S. C., Garner, J. P., Felt, S. A., Geronimo, J. T. & Chu, D. K. A "pedi" cures all: Toenail trimming and the treatment of ulcerative dermatitis in mice. *PLoS ONE* **11**, e0144871 (2016). doi: 10.1371/journal.pone.0144871

27. Gottlieb, D. H., Capitanio, J. P. & McCowan, B. Risk factors for stereotypic behavior and self-biting in rhesus macaques (*Macaca mulatta*): Animal's history, current environment, and personality. *American Journal of Primatology* **75**, 995–1008 (2013). doi: 10.1002/ajp.22161

28. Kauffman, A. S., Paul, M. J., Butler, M. P. & Zucker, I. Huddling, locomotor, and nest-building behaviors of furred and furless Siberian hamsters. *Physiology & Behavior* **79**, 247–256 (2003). doi: 10.1016/s0031-9384(03)00115-x

29. Sullivan, E. L., Koegler, F. H. & Cameron, J. L. Individual differences in physical activity are closely associated with changes in body weight in adult female rhesus monkeys (*Macaca mulatta*). *American Journal of Physiology-Regulatory Integrative and Comparative Physiology* **291**, R633–R642 (2006). doi: 10.1152/ajpregu.00069.2006

30. Rushen, J. The coping hypothesis of stereotypic behavior. *Animal Behaviour* **45**, 613–615 (1993). doi: 10.1006/anbe.1993.1071

31. Ijichi, C. L., Collins, L. M. & Elwood, R. W. Evidence for the role of personality in stereotypy predisposition. *Animal Behaviour* **85**, 1145–1151 (2013). doi: 10.1016/j.anbehav.2013.03.033

32. McBride, S. D. & Cuddeford, D. The putative welfare-reducing effects of preventing equine stereotypic behaviour. *Animal Welfare* **10**, 173–189 (2001).

33. Würbel, H. & Stauffacher, M. Prevention of stereotypy in laboratory mice: Effects on stress physiology and behaviour. *Physiology & Behavior* **59**, 1163–1170 (1996). doi: 10.1016/0031-9384(95)02268-6

34. Fureix, C. et al. Stereotypic behaviour in standard non-enriched cages is an alternative to depression-like responses in C57BL/6 mice. *Behavioural Brain Research* **305**, 186–190 (2016). doi: 10.1016/j.bbr.2016.02.005

35. Hart, P. C. et al. Analysis of abnormal repetitive behaviors in experimental animal models. In Warnick, J. E. & Kalueff, A. V. (Eds.), *Translational Neuroscience in Animal Research: Advancement, Challenges, and Research Ethics* (Hauppauge, NY: Nova Science Publishers, 2010), pp. 71–82.

36. Langen, M., Kas, M. J. H., Staal, W. G., van Engeland, H. & Durston, S. The neurobiology of repetitive behavior: Of mice.... *Neuroscience and Biobehavioral Reviews* **35**, 345–355 (2011). doi: 10.1016/j.neubiorev.2010.02.004

37. Lewis, M. H., Tanimura, Y., Lee, L. W. & Bodfish, J. W. Animal models of restricted repetitive behavior in autism. *Behavioural Brain Research* **176**, 66–74 (2007). doi: 10.1016/j.bbr.2006.08.023

38. Bauman, M. D. et al. Activation of the maternal immune system during pregnancy alters behavioral development of rhesus monkey offspring. *Biological Psychiatry* **75**, 332–341 (2014). doi: 10.1016/j.biopsych.2013.06.025

39. Vieira, G. D. T., Lossie, A. C., Lay, D. C., Radcliffe, J. S. & Garner, J. P. Preventing, treating, and predicting barbering: A fundamental role for biomarkers of oxidative stress in a mouse model of Trichotillomania. *PLoS ONE* **12**, e0175222 (2017). doi: 10.1371/journal.pone.0175222

40. George, N. M. et al. Antioxidant therapies for ulcerative dermatitis: A potential model for skin picking disorder. *PLoS ONE* **10**, e0132092 (2015). doi: 10.1371/journal.pone.0132092
41. Shively, C. A. & Willard, S. L. Behavioral and neurobiological characteristics of social stress versus depression in nonhuman primates. *Experimental Neurology* **233**, 87–94 (2012). doi: 10.1016/j.expneurol.2011.09.026
42. Kalueff, A. V. et al. Neurobiology of rodent self-grooming and its value for translational neuroscience. *Nature Reviews Neuroscience* **17**, 45–59 (2016). doi: 10.1038/nrn.2015.28
43. Garner, J. P., Weisker, S. M., Dufour, B. & Mench, J. A. Barbering (fur and whisker trimming) by laboratory mice as a model of human trichotillomania and obsessive-compulsive spectrum disorders. *Comparative Medicine* **54**, 216–224 (2004).
44. Garner, J. P. Stereotypies and other abnormal repetitive behaviors: Potential impact on validity, reliability, and replicability of scientific outcomes. *ILAR Journal* **46**, 106–117 (2005). doi: 10.1093/ilar.46.2.106
45. Poirier, C. & Bateson, M. Pacing stereotypies in laboratory rhesus macaques: Implications for animal welfare and the validity of neuroscientific findings. *Neuroscience and Biobehavioral Reviews* **83**, 508–515 (2017). doi: 10.1016/j.neubiorev.2017.09.010
46. Garner, J. P. et al. Reverse-translational biomarker validation of Abnormal Repetitive Behaviors in mice: An illustration of the 4P's modeling approach. *Behavioural Brain Research* **219**, 189–196 (2011). doi: 10.1016/j.bbr.2011.01.002
47. Burrows, E. L. & Hannan, A. J. Towards environmental construct validity in animal models of CNS disorders: Optimizing translation of preclinical studies. *CNS & Neurological Disorders-Drug Targets* **12**, 587–592 (2013).
48. Mills, D. S. Medical paradigms for the study of problem behaviour: A critical review. *Applied Animal Behaviour Science* **81**, 265–277 (2003).
49. Wiedenmayer, C. Causation of the ontogenetic development of stereotypic digging in gerbils. *Animal Behaviour* **53**, 461–470 (1997). doi: 10.1006/anbe.1996.0296
50. Weber, E., Ahloy Dallaire, J., Gaskill, B., Pritchett-Corning, K. & Garner, J. P. Aggression in group-housed laboratory mice: Why can't we solve the problem? *Lab Animal* **46**, 157–161 (2017).
51. Bauman, M. D., Toscano, J. E., Babineau, B. A., Mason, W. A. & Amaral, D. G. Emergence of stereotypies in juvenile monkeys (*Macaca mulatta*) with neonatal amygdala or hippocampus lesions. *Behavioral Neuroscience* **122**, 1005–1015 (2008). doi: 10.1037/a0012600
52. Ernst, A. M. & Smelik, P. G. Site of action of dopamine and apomorphine on compulsive gnawing behaviour in rats. *Experientia* **22**, 837 (1966). doi: 10.1007/bf01897450
53. Bechard, A. R., Bliznyuk, N. & Lewis, M. H. The development of repetitive motor behaviors in deer mice: Effects of environmental enrichment, repeated testing, and differential mediation by indirect basal ganglia pathway activation. *Developmental Psychobiology* **59**, 390–399 (2017). doi: 10.1002/dev.21503
54. Garner, J. P. Perseveration and stereotypy – Systems-level insights from clinical psychology. In Mason, G. & Rushen, J. (Eds.), *Stereotypic Animal Behaviour: Fundamentals and Applications to Welfare* (2nd edition) (Wallingford, UK: CABI, 2006), pp. 121–152.
55. Harper, L. et al. Stereotypic mice are aggressed by their cage-mates, and tend to be poor demonstrators in social learning tasks. *Animal Welfare* **24**, 463–473 (2015). doi: 10.7120/09627286.24.4.463
56. Díez-León, M. et al. Environmentally enriched male mink gain more copulations than stereotypic, barren-reared competitors. *PLoS ONE* **8**, e80494 (2013). doi: 10.1371/journal.pone.0080494
57. Clubb, R. & Vickery, S. Locomotory stereotypies in carnivores: Does pacing stem from hunting, ranging or frustrated escape?. In G. Mason & J. Rushen (Eds.), *Stereotypic Animal Behaviour: Fundamentals and Applications to Welfare* (2nd edition) (Wallingford, Oxfordshire, UK: CABI, 2006), pp. 58–85.
58. Bergeron, R., Badnell-Waters, A. J., Lambton, S. & Mason, G. Stereotypic oral behaviour in captive ungulates: Foraging, diet and gastrointestinal function. In Mason, G. & Rushen, J. (Eds.), *Stereotypic Animal Behaviour: Fundamentals and Applications to Welfare* (2nd edition) (Wallingford, UK: CABI, 2006), pp. 19–57.
59. Krohn, T. C., Ritskes-Hoitinga, J. & Svendsen, P. The effects of feeding and housing on the behaviour of the laboratory rabbit. *Laboratory Animals* **33**, 101–107 (1999). doi: 10.1258/002367799780578327
60. Talbot, S., Freire, R. & Wassens, S. Effect of captivity and management on behaviour of the domestic ferret (*Mustela putorius furo*). *Applied Animal Behaviour Science* **151**, 94–101 (2014).
61. Novak, M. A. & Bollen, K. S. Differences in the prevalence and form of abnormal behaviour across primates. In Mason, G. & Rushen, J. (Eds.), *Stereotypic Animal Behaviour: Fundamentals and Applications to Welfare* (Wallingford, UK: CABI, 2006). p. 79.

62. Lutz, C. K. A cross-species comparison of abnormal behavior in three species of singly-housed old world monkeys. *Applied Animal Behaviour Science* **199**, 52–58 (2018). doi: 10.1016/j.applanim.2017.10.010

63. Rushen, J. & Mason, G. A decade-or-more's progress in understanding stereotypic behaviour. In Mason, G. & Rushen, J. (Eds.), *Stereotypic Animal Behaviour. Fundamentals and Applications to Welfare* (Wallingford, UK: CABI, 2006), pp. 1–18.

64. Lewis, M. H., Presti, M. F., Lewis, J. B. & Turner, C. A. The neurobiology of stereotypy I: Environmental complexity. In Mason, G. & Rushen, J. (Eds.), *Stereotypic Animal Behaviour: Fundamentals and Applications to Welfare* (2nd edition) (Wallingford, UK: CABI, 2006), pp. 190–226.

65. Phillips, D. et al. Cage-induced stereotypic behaviour in laboratory mice covaries with nucleus accumbens FosB/ΔFosB expression. *Behavioural Brain Research* **301**, 238–242 (2016). doi: 10.1016/j.bbr.2015.12.035

66. Tanimura, Y., King, M. A., Williams, D. K. & Lewis, M. H. Development of repetitive behavior in a mouse model: Roles of indirect and striosomal basal ganglia pathways. *International Journal of Developmental Neuroscience* **29**, 461–467 (2011). doi: 10.1016/j.ijdevneu.2011.02.004

67. Tanimura, Y., Vaziri, S. & Lewis, M. H. Indirect basal ganglia pathway mediation of repetitive behavior: Attenuation by adenosine receptor agonists. *Behavioural Brain Research* **210**, 116–122 (2010). doi: 10.1016/j.bbr.2010.02.030

68. Wolmarans, D. W., Brand, L., Stein, D. J. & Harvey, B. H. Reappraisal of spontaneous stereotypy in the deer mouse as an animal model of obsessive-compulsive disorder (OCD): Response to escitalopram treatment and basal1 serotonin transporter (SERT) density. *Behavioural Brain Research* **256**, 545–553 (2013). doi: 10.1016/j.bbr.2013.08.049

69. Hemmings, A., Parker, M. O., Hale, S. D. & McBride, S. D. Causal and functional interpretation of mu- and delta-opioid receptor profiles in mesoaccumbens and nigrostriatal pathways of an oral stereotypy phenotype. *Behavioural Brain Research* **353**, 238–242 (2018). doi: 10.1016/j.bbr.2018.06.031

70. Díez-León, M. et al. Neurophysiological correlates of stereotypic behaviour in a model carnivore species. *Behavioural Brain Research* **373**, 112056 (2019). doi: 10.1016/j.bbr.2019.112056

71. Lutz, C. K. & Novak, M. A. Environmental enrichment for nonhuman primates: Theory and application. *ILAR Journal* **46**, 178–191 (2005). doi: 10.1093/ilar.46.2.178

72. Prescott, M. J., Nixon, M. E., Farningham, D. A. H., Naiken, S. & Griffiths, M. A. Laboratory macaques: When to wean? *Applied Animal Behaviour Science* **137**, 194–207 (2012). doi: 10.1016/j.applanim.2011.11.001

73. Zhang, B. Consequences of early adverse rearing experience (EARE) on development: Insights from non-human primate studies. *Zoological Research* **38**, 7–35 (2017). doi: 10.13918/j.issn.2095-8137.2017.002

74. Latham, N. R. & Mason, G. J. Maternal deprivation and the development of stereotypic behaviour. *Applied Animal Behaviour Science* **110**, 84–108 (2008). doi: 10.1016/j.applanim.2007.03.026

75. Gharib, H. S. A., Abdel-Fattah, A. F., Mohammed, H. A. & Abdel-Fattah, D. M. Weaning induces changes in behavior and stress indicators in young New Zealand rabbits. *Journal of Advanced Veterinary and Animal Research* **5**, 166–172 (2018). doi: 10.5455/javar.2018.e262

76. Goto, A., Arata, S., Kiyokawa, Y., Takeuchi, Y. & Mori, Y. Risk factors for canine tail chasing behaviour in Japan. *Veterinary Journal* **192**, 445–448 (2012). doi: 10.1016/j.tvjl.2011.09.004

77. Ahola, M. K., Vapalahti, K. & Lohi, H. Early weaning increases aggression and stereotypic behaviour in cats. *Scientific Reports* **7**, 10412 (2017). doi: 10.1038/s41598-017-11173-5

78. Richter, S. H., Kastner, N., Loddenkemper, D. H., Kaiser, S. & Sachser, N. A time to wean? Impact of weaning age on anxiety-like behaviour and stability of behavioural traits in full adulthood. *PLoS ONE* **11**, e0167652 (2016). doi: 10.1371/journal.pone.0167652

79. Latham, N. & Mason, G. From house mouse to mouse house: The behavioural biology of free-living *Mus musculus* and its implications in the laboratory. *Applied Animal Behaviour Science* **86**, 261–289 (2004). doi: 10.1016/j.applanim.2004.02.006

80. Lehmann, M. Social-behavior in young domestic rabbits under seminatural conditions. *Applied Animal Behaviour Science* **32**, 269–292 (1991). doi: 10.1016/s0168-1591(05)80049-8

81. Bradshaw, J. W. S., Casey, R. & Brown, S. *The Behaviour of the Domestic Cat* (Wallingford, UK: CABI, 2012).

82. Paul, M. & Bhadra, A. Selfish pups: Weaning conflict and milk theft in free-ranging dogs. *PLoS ONE* **12**, e0170590 (2017). doi: 10.1371/journal.pone.0170590

83. Bauer, S. A. & Baker, K. C. Persistent effects of peer rearing on abnormal and species-appropriate activities but not social behavior in group-housed rhesus macaques (*Macaca mulatta*). *Comparative Medicine* **66**, 129–136 (2016).

84. Würbel, H. Ideal homes? Housing effects on rodent brain and behaviour. *Trends in Neurosciences* **24**, 207–211 (2001). doi: 10.1016/s0166-2236(00)01718-5

85. Lewis, M. H. Environmental complexity and central nervous system development and function. *Mental Retardation and Developmental Disabilities Research Reviews* **10**, 91–95 (2004). doi: 10.1002/mrdd.20017

86. Latham, N. & Mason, G. Frustration and perseveration in stereotypic captive animals: Is a taste of enrichment worse than none at all? *Behavioural Brain Research* **211**, 96–104 (2010). doi: 10.1016/j.bbr.2010.03.018

87. Gross, A. N., Richter, S. H., Engel, A. K. J. & Wurbel, H. Cage-induced stereotypies, perseveration and the effects of environmental enrichment in laboratory mice. *Behavioural Brain Research* **234**, 61–68 (2012). doi: 10.1016/j.bbr.2012.06.007

88. Appleby, M. C., Lawrence, A. B. & Illius, A. W. Influence of neighbors on stereotypic behavior of tethered sows. *Applied Animal Behaviour Science* **24**, 137–146 (1989). doi: 10.1016/0168-1591(89)90041-5

89. Hennessy, M. B., McCowan, B., Jiang, J. & Capitanio, J. P. Depressive-like behavioral response of adult male rhesus monkeys during routine animal husbandry procedure. *Frontiers in Behavioral Neuroscience* **8**, 309 (2014). doi: 10.3389/fnbeh.2014.00309

90. Vandeleest, J. J., McCowan, B. & Capitanio, J. P. Early rearing interacts with temperament and housing to influence the risk for motor stereotypy in rhesus monkeys (*Macaca mulatta*). *Applied Animal Behaviour Science* **132**, 81–89 (2011). doi: 10.1016/j.applanim.2011.02.010.

91. Lutz, C., Well, A. & Novak, M. Stereotypic and self-injurious behavior in rhesus macaques: A survey and retrospective analysis of environment and early experience. *American Journal of Primatology* **60**, 1–15 (2003), doi: 10.1002/ajp.10075

92. Gottlieb, D. H., Maier, A. & Coleman, K. Evaluation of environmental and intrinsic factors that contribute to stereotypic behavior in captive rhesus macaques (*Macaca mulatta*). *Applied Animal Behaviour Science* **171**, 184–191 (2015). doi: 10.1016/j.applanim.2015.08.005

93. Lewis, R. S. & Hurst, J. L. The assessment of bar chewing as an escape behaviour in laboratory mice. *Animal Welfare* **13**, 19–25 (2004).

94. Nevison, C. M., Hurst, J. L. & Barnard, C. J. Why do male ICR(CD-1) mice perform bar-related (stereotypic) behaviour? *Behavioural Processes* **47**, 95–111 (1999). doi: 10.1016/s0376-6357(99)00053-4

95. NRC. *Guide for the Care and Use of Laboratory Animals* (Washington, DC: The National Academies Press, 2011).

96. European Union *Directive 2010/63: Legislation for the Protection of Animals Used for Scientific Purposes* (2010).

97. Coleman, K. & Maier, A. The use of positive reinforcement training to reduce stereotypic behavior in rhesus macaques. *Applied Animal Behaviour Science* **124**, 142–148 (2010). doi: 10.1016/j.applanim.2010.02.008

98. Simpson, J. & Kelly, J. P. The impact of environmental enrichment in laboratory rats: Behavioural and neurochemical aspects. *Behavioural Brain Research* **222**, 246–264 (2011). doi: 10.1016/j.bbr.2011.04.002

99. Bailoo, J. D. et al. Effects of cage enrichment on behavior, welfare and outcome variability in female mice. *Frontiers in Behavioral Neuroscience* **12**, 232 (2018). doi: 10.3389/fnbeh.2018.00232

100. Baumans, V. Environmental enrichment for laboratory rodents and rabbits: Requirements of rodents, rabbits, and research. *ILAR Journal* **46**, 162–170 (2005).

101. Coleman, K. & Novak, M. A. Environmental enrichment in the 21st century. *Ilar Journal* **58**, 295–307 (2017). doi: 10.1093/ilar/ilx008

102. Burn., C. C., Raffle, J. & Bizley, J. K. Does "playtime" reduce stimulus-seeking and other boredom-like behaviour in laboratory ferrets? *Animal Welfare* **29**, 19–26 (2020). doi: 10.7120/09627286.29.1.019

103. Ellis, S. Environmental enrichment: Practical strategies for improving feline welfare. *Journal of Feline Medicine and Surgery* **11**, 901–912 (2009). doi: 10.1016/j.jfms.2009.09.011

104. Fontenot, M. B. et al. The effects of fluoxetine and buspirone on self-injurious and stereotypic behavior in adult male rhesus macaques. *Comparative Medicine* **55**, 67–74 (2005).

105. Fontenot, M. B., Musso, M. W., McFatter, R. M. & Anderson, G. M. Dose-finding study of fluoxetine and venlafaxine for the treatment of self-injurious and stereotypic behavior in rhesus macaques (*Macaca mulatta*). *Journal of the American Association for Laboratory Animal Science* **48**, 176–184 (2009).

106. Poulsen, E. M. B., Honeyman, V., Valentine, P. A. & Teskey, G. C. Use of fluoxetine for the treatment of stereotypical pacing behavior in a captive polar bear. *Journal of the American Veterinary Medical Association* **209**, 1470–1474 (1996).

107. Bayne, K. et al. The use of artificial turf as a foraging substrate for individually housed rhesus monkeys (*Macaca mulatta*). *Animal Welfare* **1**, 39–53 (1992).

108. Novak, M. A., Kinsey, J. H., Jorgensen, M. J. & Hazen, T. J. Effects of puzzle feeders on pathological behavior in individually housed rhesus monkeys. *American Journal of Primatology* **46**, 213–227 (1998). doi: 10.1002/(sici)1098-2345(1998)46:3<213::Aid-ajp3>3.0.Co;2-l

109. Bourgeois, S. R. & Brent, L. Modifying the behaviour of singly caged baboons: Evaluating the effectiveness of four enrichment techniques. *Animal Welfare* **14**, 71–81 (2005).

110. Polanco, A., Díez-León, M. & Mason, G. Stereotypic behaviours are heterogeneous in their triggers and treatments in the American mink, *Neovison vison*, a model carnivore. *Animal Behaviour* **141**, 105–114 (2018). doi: 10.1016/j.anbehav.2018.05.006

111. Rommeck, I., Anderson, K., Heagerty, A., Cameron, A. & McCowan, B. Risk factors and remediation of self-injurious and self-abuse behavior in rhesus macaques. *Journal of Applied Animal Welfare Science* **12**, 61–72 (2009). doi: 10.1080/10888700802536798

112. Novak, J., Bailoo, J. D., Melotti, L. & Würbel, H. Effect of cage-induced stereotypies on measures of affective state and recurrent perseveration in CD-1 and C57BL/6 mice. *PLoS ONE* **11**, e0153203 (2016). doi: 10.1371/journal.pone.0153203

113. Powell, S. B., Newman, H. A., McDonald, T. A., Bugenhagen, P. & Lewis, M. H. Development of spontaneous stereotyped behavior in deer mice: Effects of early and late exposure to a more complex environment. *Developmental Psychobiology* **37**, 100–108 (2000). doi: 10.1002/1098-2302(200009)37:2<100::Aid-dev5>3.0.Co;2-6

114. Zabegalov, K. N. et al. Abnormal repetitive behaviors in zebrafish and their relevance to human brain disorders. *Behavioural Brain Research* **367**, 101–110 (2019). doi: 10.1016/j.bbr.2019.03.044

115. Kistler, C., Hegglin, D., Würbel, H. & Konig, B. Preference for structured environment in zebrafish (*Danio rerio*) and checker barbs (*Puntius oligolepis*). *Applied Animal Behaviour Science* **135**, 318–327 (2011). doi: 10.1016/j.applanim.2011.10.014

116. Huntingford, F. A. & Kadri, S. Defining, assessing and promoting the welfare of farmed fish. *Revue Scientifique Et Technique-Office International Des Epizooties* **33**, 233–244 (2014). doi: 10.20506/rst.33.1.2286

117. Hubrecht, R. C. Enrichment in puppyhood and its effects on later behavior of dogs. *Laboratory Animal Science* **45**, 70–75 (1995).

118. Hubrecht, R. C., Serpell, J. A. & Poole, T. B. Correlates of pen size and housing conditions on the behavior of kenneled dogs. *Applied Animal Behaviour Science* **34**, 365–383 (1992). doi: 10.1016/s0168-1591(05)80096-6

119. Novak, J., Bailoo, J. D., Melotti, L., Rommen, J. & Würbel, H. An exploration based cognitive bias test for mice: Effects of handling method and stereotypic behaviour. *PLoS ONE* **10**, e0130718 (2015). doi: 10.1371/journal.pone.0130718

120. Muehlmann, A. M. et al. Further characterization of repetitive behavior in C58 mice: Developmental trajectory and effects of environmental enrichment. *Behavioural Brain Research* **235**, 143–149 (2012). doi: 10.1016/j.bbr.2012.07.041

121. Callard, M. D., Bursten, S. N. & Price, E. O. Repetitive backflipping behaviour in captive roof rats (*Rattus rattus*) and the effects of cage enrichment. *Animal Welfare* **9**, 139–152 (2000).

122. Gunn, D. & Morton, D. B. Inventory of the behavior of New Zealand white rabbits in laboratory cages. *Applied Animal Behaviour Science* **45**, 277–292 (1995). doi: 10.1016/0168-1591(95)00627-5

123. Chu, L. R., Garner, J. P. & Mench, J. A. A behavioral comparison of New Zealand White rabbits (*Oryctolagus cuniculus*) housed individually or in pairs in conventional laboratory cages. *Applied Animal Behaviour Science* **85**, 121–139 (2004). doi: 10.1016/j.applanim.2003.09.011

124. Vieuille-Thomas, C., Le Pape, G. & Signoret, J. P. Stereotypies in pregnant sows: Indications of influence of the housing system on the patterns expressed by the animals. *Applied Animal Behaviour Science* **44**, 19–27 (1995). doi: 10.1016/0168-1591(95)00574-c

125. Bellanca, R. U. & Crockett, C. M. Factors predicting increased incidence of abnormal behavior in male pigtailed macaques. *American Journal of Primatology* **58**, 57–69 (2002). doi: 10.1002/ajp.10052

126. Doyle, L. A., Baker, K. C. & Cox, L. D. Physiological and behavioral effects of social introduction on adult male rhesus macaques. *American Journal of Primatology* **70**, 542–550 (2008). doi: 10.1002/ajp.20526

127. Sarna, J. R., Dyck, R. H. & Whishaw, I. Q. The Dalila effect: C57BL6 mice barber whiskers by plucking. *Behavioural Brain Research* **108**, 39–45 (2000). doi: 10.1016/s0166-4328(99)00137-0

128. Garner, J. P. & Mason, G. J. Evidence for a relationship between cage stereotypies and behavioural disinhibition in laboratory rodents. *Behavioural Brain Research* **136**, 83–92 (2002). doi: 10.1016/s0166-4328(02)00111-0

129. Pinelli, C. J., Leri, F. & Turner, P. V. Long term physiologic and behavioural effects of housing density and environmental resource provision for adult male and female Sprague Dawley rats. *Animals* **7**, 44 (2017). doi: 10.3390/ani7060044

130. Podberscek, A. L., Blackshaw, J. K. & Beattie, A. W. The behavior of group penned and individually caged laboratory rabbits. *Applied Animal Behaviour Science* **28**, 353–363 (1991). doi: 10.1016/0168-1591(91)90167-v

131. Hurst, J. L., Barnard, C. J., Tolladay, U., Nevison, C. M. & West, C. D. Housing and welfare in laboratory rats: Effects of cage stocking density and behavioural predictors of welfare. *Animal Behaviour* **58**, 563–586 (1999). doi: 10.1006/anbe.1999.1165

132. Abou-Ismail, U. A., Burman, O. H. P., Nicol, C. J. & Mendl, M. The effects of enhancing cage complexity on the behaviour and welfare of laboratory rats. *Behavioural Processes* **85**, 172–180 (2010). doi: 10.1016/j.beproc.2010.07.002

133. Abou-Ismail, U. A. & Mendl, M. T. The effects of enrichment novelty versus complexity in cages of group-housed rats (*Rattus norvegicus*). *Applied Animal Behaviour Science* **180**, 130–139 (2016). doi: 10.1016/j.applanim.2016.04.014

134. Hennessy, M. B., Chun, K. & Capitanio, J. P. Depressive-like behavior, its sensitization, social buffering, and altered cytokine responses in rhesus macaques moved from outdoor social groups to indoor housing. *Social Neuroscience* **12**, 65–75 (2017). doi: 10.1080/17470919.2016.1145595

135. Qin, D. et al. A spontaneous depressive pattern in adult female rhesus macaques. *Scientific Reports* **5**, 11267 (2015). doi: 10.1038/srep11267

5 Animal Learning
The Science behind Animal Training

Dorte Bratbo Sørensen, Annette Pedersen, and Björn Forkman

CONTENTS

LEARNING THEORIES

The evolutionary function of learning is for an animal to be able to predict and control her/his environment. Learning requires that the animal recognizes changes in the environment (i.e., perceives the relevant stimuli) and is able to recall the event and – depending on what has been learned – her/his own behavior (i.e., the behavior before the event). Animals learn all the time. To avoid our animals developing fear responses towards the surroundings or us, we need to know the principles behind animal learning. Using the principles described in this chapter, it is also possible to adapt the environment and our own behavior to ensure – by determinedly focused handling, training, and management procedures – that the animal mostly will learn responses and behaviors that increase the welfare of each individual animal.

In this chapter, we will thus use the word "learning" to denote the behavioral changes shown by an animal because of experience. The neurochemistry and neuroplasticity underlying learning and memory are outside the scope of this chapter.

STIMULI AND GENERALIZATION

A prerequisite for learning is the ability to perceive stimuli, both external and internal. An external stimulus may be a physical change such as a change in wavelength of light in the environment or a sound. An internal stimulus could be a change in heart rate or the experience of nausea. A stimulus is only considered a stimulus if the animal is able to perceive it. Hence, the animal must possess the sensory capacity to detect the stimulus. Individual animals of the same species may have diverging abilities to detect a specific stimulus. The sensory capacity to detect a certain stimulus differs between species and may differ between individuals. The sound of rats vocalizing at 50 kHz (in the ultrasonic range) is not a stimulus that affects human behavior, as we are not able to detect ultrasound. To other rats, the sound emitted by the vocalizing rat may hold important information and the animals may choose to respond to the stimulus.

When an animal encounters a new stimulus that shares characteristics with a known stimulus and responds in a similar way, the animal is demonstrating stimulus generalization. From an evolutionary perspective, generalization is very important. If, for example, an animal has learned that a certain sound in the undergrowth (e.g., the cracking of a dead branch close by) is followed by the rapid attack from a predator, it is purposeful that the animal – after a narrow escape – will react to sounds that share characteristics with the previous sound. Even though a new sound is not exactly the same or not encountered in the exact same context, the animal will still react as though it could signal a predator hiding in the undergrowth.

In the laboratory facility stimulus generalization also happens. Consider a pig that has learned to enter a weighing wagon and get an apple when the caretaker stands next to the wagon facing the pig and waves her/his hand. The pig will then generalize to other people standing close to the wagon – perhaps in a slightly different position – moving their hands with an almost similar waving-like motion. The more the pig is able to generalize, the more she/he will respond to sloppy signaling from the trainer. If – on the other hand – the trainer wants the pig to show two different behavioral responses to two almost identical hand signals, then the trainer does not want the pig to generalize all hand movements to indicate one response. Instead, the pig must learn to discriminate between the almost identical hand signals in order to be able to show the correct response for each hand signal. Whether an animal may learn to either discriminate between stimuli or generalize, depends on a large variety of factors.

When an animal learns an association between her/his own actions and a resulting consequence, very soon the animal will also associate certain stimuli in the surroundings with whether or not the behavior will have the expected consequences. For example, the squirrel learns that jumping to the bird table only will be worth the effort after the Lady-of-the-patio-door has been standing close to the bird table and called "birdie-birdie-birdie." The sight of the Lady and the sound of her "birdie call" are stimuli telling the squirrel that new food (including kernels of hazelnut and peanuts) is now on the table and the jumping behavior onto the bird table will be worthwhile. At other times, when no calling has been heard, there will be none of the good stuff on the bird table. Hence, jumping behavior only pays off when the stimulus "birdie-birdie-birdie" is present prior to the performance of the behavior. Such a stimulus is called a discriminative stimulus (S^d), as it helps the squirrel discriminate between different situations with different opportunities for obtaining highly valued kinds of food. An S^d is in training situations also known as a "cue" or a "signal."

Stimulus control is an animal training concept related to generalization and discrimination. When an animal is trained using operant conditioning techniques, the trainer would like the animal to discriminate the trainer's signal from all other stimuli and respond immediately showing the correct behavior every time the specific signal is given. No other behavioral responses should be shown. In addition, this particular behavioral response should only be shown when the trainer has given the signal and not on other signals or in other contexts. Getting a behavior under stimulus control hence demands that the animal can clearly discriminate the trainer's signal from other stimuli. On the other hand, the ability of the animal to generalize is also relevant. For example, the trainer may like the animal to respond when other trainers give the signal or when the signal is given in slightly different surroundings.

LEARNING

Learning is traditionally classified as either non-associative learning or associative learning. When an animal learns something about the environment without the perceived stimulus being associated with other events or the individual's own behavior, it is called non-associative learning. Habituation is an example of non-associative learning. When two or more events are associated, so that one stimulus or event (or a specific combination of stimuli) predicts another event or stimulus, it is called associative learning. The same is the case when an animal learns to associate her/his own actions with the consequences of that particular action.

NON-ASSOCIATIVE LEARNING

SENSITIZATION AND HABITUATION

Encountering a new event, e.g., a predator, will create an expectation of more of the same. This expectation makes the animal more sensitive to all changes in her/his environment (irrespective of whether they are olfactory, auditory, or visual), and the animal will react to all of these as if each change predicted the expected event, in this case a predator. The same mechanism is true for pleasant events, e.g., if finding new food, the animal will form an expectation of more food and will interpret all events as predicting new food. If the animal does not encounter new predators (or food) then the expectancy, and thus the sensitization gradually disappears over time.

Habituation is defined as decreased responding to repeated presentation of a specific stimulus. It is stimulus specific, in contrast to sensitization that is not specific to the stimuli eliciting the response. Habituation causes a gradual decrease in the response initially elicited by the presentation of a new stimulus. Thus, a behavioral adaptability ensures that the animal will not waste energy on responding to stimuli of no importance or interest. Take, for example, a guinea pig encountering a short, abrupt noise from a thermostat in the housing room. The guinea pig immediately takes cover and will be more alert. However, nothing happens. This sound will appear from time to time, when the thermostat turns on and off, and still nothing ever happens following the sound. Over time, the flight response of the guinea pig will cease to be elicited by the sound.

Both habituation and sensitization may be activated, when a stimulus is presented and often an initial sensitization is seen prior to the development of habituation (Figure 5.1).[1]

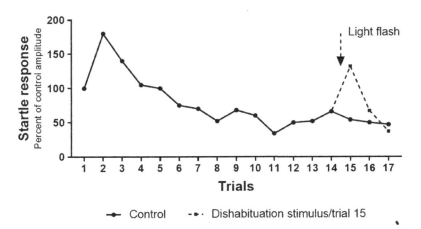

The independence of habituation and sensitization
(modified from Groves and Thompson, 1970)

FIGURE 5.1 Rats exposed to a total of 17 habituation stimuli (a 110 dB, 1000 Hz tone for two seconds) will show an initially increased startle response (sensitization) prior to habituation (the control curve). If a stimulus (one second of flashing light) is applied just prior to trial no.15 to some of the animals, an increase in the response to the tone, a dishabituation, will be shown (dotted line) compared to control animals that are only subjected to the tone and not the additional flashing light. (Modified from Groves and Thompson, 1970)

It is important to realize that habituation is not muscular fatigue. The guinea pig does not stop responding to the sound due to exhaustion; if a different, sudden noise is presented, the new stimulus will still make the guinea pig run for shelter.

Desensitization is a special case of habituation used as a behavioral modification technique when animals are housed in an environment with potentially fear-eliciting stimuli such as the sound of an electric clipper. If such a device is turned on right next to the animal, most animals will be startled and try to get away. During desensitization, the eliciting stimulus (e.g., the electric clipper) is kept at such a low intensity that it only elicits a very low-level behavioral or emotional response. Instead of just turning on the electric clipper in front of the animal, the clipper is turned on, e.g., in the next room, and the behavior of the animal is closely monitored. If the animal does not react, the clipper is gradually brought closer and closer to the animal, making sure that the animal shows no signs of fear. Done correctly, it will – after desensitization – be possible to turn on the clipper right in front of the animal without eliciting signs of fear. The advantage of this technique is that – when used correctly – there is only a very low risk of increased fear response from the animal.

On a practical level, the phenomena of habituation and sensitization are important to bear in mind when transporting and introducing laboratory animals to a new facility. It is well-established that the animals must be allowed time to recover from the stress of transportation.[2] It is reasonable to suggest that the transportation has sensitized the animal and therefore that the new environment may hold a large portion of potentially frightening (but not dangerous) stimuli for which the animal should be allowed proper habituation time.

ASSOCIATIVE LEARNING

There are two types of associative learning, classical conditioning and operant conditioning. In classical conditioning, an association between two stimuli is learned – with one stimulus predicting the other. Classical conditioning is also called Pavlovian conditioning. In operant conditioning, the animal associates her/his own behavior with the consequence of that particular behavior. Operant conditioning is also called instrumental conditioning or sometimes trial and error learning.

CLASSICAL CONDITIONING

To be successful in this world, being able to predict danger is an important qualification. Moreover, to be able to predict the appearance of good things and where to find them, is just as important. If a squirrel has associated the sound of a little bell with the presence of the neighborhood cat, he can escape before the cat gets close enough to attack. If he has associated the sound of the opening of the patio door next to the large pine with the presence of grains, seed, and nuts on the bird table, then he can get all the good stuff before the pigeons and magpies arrive. Learning to associate one event with another event is known as classical conditioning.

The principles of classical conditioning were originally pinpointed by Ivan Pavlov, who was a Russian physiologist. He worked with dogs from which he collected saliva when the dogs were fed. The food represented an unconditioned ("un-learned") stimulus (US) that the dogs innately reacted to by salivating (an "un-learned"/unconditioned response; UR). The dogs began to associate certain stimuli (e.g., the sound of Pavlov's footsteps or the smell of food) with the arrival of the food. Soon the dogs began to salivate on the presentation of these specific stimuli, even though no food was in sight. Pavlov realized that a variety of stimuli could be conditioned and thus trigger salivation in his dogs prior to feeding. In the most well-known example, a bell was sounded just prior to food delivery and after having presented first the sound of the bell and then the food a number of times, the dogs began to associate the bell with the food. As a result, the dogs started salivating as soon as they heard the sound of the bell, even though the food was not yet presented to the dogs.

Through classical conditioning, animals learn an association between two stimuli (or events). In Pavlov's classical experiment a bell was rung before food was presented to the dog. After a number of repetitions the reaction of the dog to the bell was the same as the reaction to the food – the dog started salivating. The reaction to the bell (the conditioned response (CR)) to the conditioned stimulus (CS) is identical to the innate reaction (UR) to the food (US). It is important to remember that the dog has not learnt to salivate when she/he hears the bell; the salivation is not an intentional response to obtain food. The dog has merely formed an association between the bell and the food and it is that association that causes the behavior.

In general, there are two major characteristics of classical conditioning. One characteristic is that both of the stimuli appear independent of the behavior of the animal, i.e., the animal has no control over the events. The other characteristic is that the behavior elicited by the US is an innate response and hence not a deliberate, conscious, intentional behavioral response.

It is important to emphasize that the UR is not necessarily a physical behavior such as a startle or an eye blink. Physiological responses may also be classically conditioned. For example, the release of specific hormones in male rats (UR) following the scent of a sexually receptive female rat (US) may be conditioned by pairing the scent of the female rat with the smell of methyl salicylate. After conditioning, the smell of methyl salicylate (CS) alone may induce the release of the hormones.[3] Another example is that the taste of saccharine-sweetened water (CS) after pairing with the immunosuppressive drug cyclosporine A (US) results in inhibition of interleukin (IL)-2 and interferon (IFN)-c production[4] in conditioned animals given saccharine-sweetened water.

Emotional responses such as fear and aggression may also be conditioned and are known as conditioned emotional responses (CER). Fear conditioning, which is a classical type of CER, is often used in research as a procedure by which an animal learns to associate, e.g., a sound, with an aversive stimulus such as an electric shock. The immediate physical response to the shock – and later the sound – is startle, freezing, and attempting to escape. Interestingly, the presentation of the CS (the sound) that has become associated with the aversive stimulus (the shock) will suppress a previously positively reinforced learned behavior.[5] This phenomenon has been extensively studied in rats in Skinner boxes (see Figure 6.1), where rats were trained to press a lever to obtain a food reinforcer. When the rats had learned this behavioral response, the researchers would then introduce a sound or a light signal (CS) paired with a brief shock (US). The rats soon associated the CS with the shock, which they feared and this conditioned fear resulted in a suppression of the lever pressing during the presentation of the CS.[5] Another example could be a pig that has learned a behavior (for example "follow target"). The pig is performing well, follows the target reliably whenever the trainer presents the target. One day, the local veterinarian (wearing a white lab coat as opposed to the trainers who all wear green clothes) comes into the pen without the trainer knowing about it, captures and fixates the pig and draws a blood sample. The blood sampling is repeated over the next days and the pig soon associates the lab coat with fixation and pain. The next week, the trainer wants to demonstrate the target behavior to the veterinarian. The pig performs nicely until the veterinarian enters the room. The sight of the white lab coat (CS) elicits a CER, namely fear, in the pig and this state suppresses the previously learned target behavior. The consequence is that while the veterinarian is present, the pig will not do the target behavior.

In animal training, the conditioned emotional response will be affected by two factors; the intensity of the applied aversive stimulus and the reinforcement history on the positively reinforced learned behavior. If an animal has been trained to stand still for an injection ("standing still") with saline and then is injected with a drug that stings, the pain may result in an emotional reaction. Whether this pain will suppress the learned "standing still" response depends on how aversive the stinging was and how well established the trained behavior is. If the animal has had 100 successful injections with saline, one event with a stinging drug will most likely not suppress the behavior seriously in the future.

Even though most research has been done on – and made use of – conditioning of negative emotional responses, positive emotional states such as anticipation and joy can also be conditioned. The conditioning of a positive emotional state is used in counter-conditioning; a behavioral modification technique used for situations where an animal experiences stress or fear induced by a specific stimulus. Through counter-conditioning, the animal is trained to associate a positive event such as the delivery of a treat with the stimulus or situation that previously induced stress or fear.

The most effective classical conditioning takes place when the CS is presented immediately (i.e., less than a few seconds) before the US. This condition holds for most cases; however, there are exceptions. One of these is taste aversion.

The phenomenon of taste aversion was studied by, among others, John Garcia and Robert Koelling,[6] whose research helped to explain how "bait shyness" arises in wild rats. Bait shyness is the behavior that emerges in wild rats having eaten poisoned food in an amount high enough to make them ill, but not high enough to kill them. Consequently, the rat will avoid this particular taste in the future. In taste aversion learning, up to six hours may elapse between the presentation of the CS (smell/flavor of an ingested food item) and the following US (illness, nausea, or other forms of physical discomfort), yet the animal still learns the connection.[7,8] One of the first studies done by Garcia and his coworkers demonstrated that when given a choice between tap water and saccharine water, rats prefer the sweet saccharine water. The sweet water was then offered to the rats coinciding with exposure to gamma radiation, which most likely made the rats nauseous. As a result, the rats avoided the sweet water for up to 30 days after irradiation.[9] It is worth noting that taste aversion can be learned with just one pairing of a specific flavor and the resulting illness. The long-delay learning of a taste aversion has most likely evolved to protect animals and humans against repeatedly eating harmful foods. Garcia also demonstrated that taste aversion depended on the nature of the CS. Rats would associate lithium-induced nausea (US) with a previously encountered gustatory stimulus (CS) such as sweet or salty water ("tasty water") and subsequently avoid the water. If, however, the drinking spout mechanism elicited sound and light (an audio-visual CS, which Garcia named "noisy-bright" water) when the rat was drinking the water and this drinking-response was then followed by a dose of lithium (US), the rats did not develop an avoidance toward the water. Hence, an association between sound/light and nausea is not as easily established. Interestingly enough, if the noisy-bright water (i.e., sound and light) was followed by electric shock, the rats would subsequently avoid the water.[6] Evidently, some associations such as the association between a new, certain taste and subsequently feeling ill or the association between environmental noise and light and getting an electric shock are more easily formed than other associations.

THE EFFECT OF SALIENCE AND CONTEXT

The speed of the conditioning is affected by the salience of the CS. The stronger the stimulus or the more it deviates from the environment, the stronger the salience, and the faster the conditioning. If the environment is very noisy, an auditory signal needs to be very loud to be conditioned or if the environment is filled with many different lights, a light signal needs to be very different for it to be effective. Likewise, new stimuli or rare stimuli will have a stronger salience than stimuli that the animal regularly encounters. It is, however, only the speed of the conditioning that is affected. If given enough pairings the strength of the conditioned response will be independent of the salience of the CS.

Animals will associate all stimuli present in the learning situation with the US. The strength of the association between the CS and the US is made up of the combination of the different stimuli that are present. This means that if an animal is trained to respond to a given sound, say a clicker, in one environment she/he may not respond to the same sound in a novel environment. This is because none of the other stimuli from the old environment, apart from the sound of the clicker, are now present. To get the animal to respond in a novel environment it is important to have trained the behavior in as many different environments as possible. It must also be noted that if a stimulus is encountered repeatedly, the animal will habituate and soon cease to respond. It could be that the kids have been playing with the clicker all morning prior to the first training session with the puppy. The puppy has now learned that the sound of the clicker is just noise that does not predict any events. This learned non-contingency may result in slower acquisition if a conditioning is attempted either in the same environment or in another environment (e.g., at puppy class training). In the same way, if rats have learned that a sound in a Skinner box is randomly presented and not associated with the shocks also presented in this particular box, it may be more difficult to establish a fear-conditioning with that sound in another environment.[10] This phenomenon is also known as latent inhibition.[11]

EXTINCTION IN CLASSICAL CONDITIONING

If the conditioned stimulus is presented repeatedly without being followed by the unconditioned stimulus, a reduction in the conditioned response will happen. In other words, if Pavlov rang the bell a number of times without feeding the dogs, the dogs would stop salivating. The phenomenon is called Extinction and it is simply the waning of a conditioned response (CR) when the US ceases to appear. Extinction is not the same as forgetting. An animal will demonstrate spontaneous recovery, i.e., a partial return of the response after a certain time. If, for example, a CS has been presented repeatedly without the US to a degree where the animal has stopped responding, extinction has taken place. If, however, the animal is presented the CS the next day, the CR may partly return. Therefore, the animal has not forgotten the response and the association, but has rather learned that there is no longer an association. Extinction has also been shown to occur when the association between the US and the CS is broken simply by offering the US without it being contingent on the CS.[12]

OPERANT CONDITIONING

When the strengthening or weakening of a behavior is a function of the consequences of that behavior, the type of learning involved is referred to as operant learning (or instrumental learning). The two most prominent contributors are Edward E. Thorndike and B.F. Skinner. Thorndike phrased The Law of Effect:

> Of several responses made to the same situation, those which are accompanied or closely followed by satisfaction to the animal will, other things being equal, be more firmly connected with the situation, so that, when it recurs, they will be more likely to recur; those which are accompanied or closely followed by discomfort to the animal will, other things being equal, have their connections with that situation weakened, so that, when it recurs, they will be less likely to occur. The greater the satisfaction or discomfort, the greater the strengthening or weakening of the bond.[13]

Skinner pioneered the development of behavior analysis and created the operant chamber known as the Skinner box.

One of the key features in operant conditioning is that events occur as a consequence of a behavior. Thorndike referred to these consequences as "a satisfying state of affairs" (e.g., the world becomes a better place) and an "annoying state of affairs." which is the opposite, as the world becomes less nice. These two kinds of consequences are now referred to as Reinforcers and Aversives (or Punishers) and are – simply put – stimuli that animals will seek/approach/work for (reinforcers) and stimuli they will try to escape and/or avoid (aversives). Reinforcers as well as aversives may be simple stimuli that are rewarding or aversive in themselves, such as a meatball, a ray of sun, or an electric shock. However, reinforcers and aversives may also be several stimuli combined to make up an event, a specific context or – in the case of reinforcers – an opportunity to obtain further reinforcers. It may also be objects that can be used to express a preferred behavior such as play behavior.

Everything we do has consequences. These consequences can be pictured as a continuum from extremely pleasurable to highly aversive. When a mouse runs across a garden path, the consequence of running across the path can be really bad if the mouse is spotted by a cat. However, if the mouse runs across the garden path and reaches a receptive female mouse on the other side, the consequences of running across the path may be highly rewarding. When a behavioral response is followed by something pleasant, i.e., a reinforcer that improves the life of the animal expressing the behavior, the animal will be more inclined to perform that behavior the next time an opportunity arises. It is important to remember that good consequences may be either obtaining something nice or the disappearance of unpleasant stimuli from the environment. For example, to a horse getting access to pasture or a piece of carrot is nice and rewarding, the release of an unpleasant pressure of the bridle by the rider is a relief that most likely is also perceived as a pleasant event from the horse's perspective.

If, on the other hand, a behavior is followed by something aversive, the animal will be less inclined to repeat that particular behavior. An aversive event may be either the addition of something unpleasant or the removal of something pleasant (or even removal of the mere opportunity to obtain something pleasant). None of these consequences are favorable from the animals' point-of-view. The aversive consequence will reduce the probability of the preceding behavior being repeated. Moreover – and importantly – the animal may also come to associate the aversive stimulus with the emotional response elicited by the stimulus and perhaps with an unpredictable number of other stimuli in the environment. If a trainer uses positive punishment (e.g., grabbing and slightly twisting the ears of a puppy to stop him from growling), the emotional response of pain and fear may be conditioned to the proximity of the trainer and the sight of the trainer will from now on induce fear in the puppy. Additionally some of the smells and sounds in the room where "ear-twisting" happens, may become cues that will elicit the same emotional response in the puppy in the future.

In operant condition learning, we hence have four scenarios that will either result in a behavior being strengthened (i.e., reinforced) or weakened (i.e., punished): the addition of a stimulus, which is either pleasurable or aversive, or the removal of either a pleasant or an aversive stimulus (Figure 5.2). When the consequence/stimulus is added to the situation, it is called positive (as in the mathematical symbol "+") and when it is removed ("subtracted") from the situation, it is called negative (as in the mathematical symbol "−"). That results in four possible scenarios that can follow a behavior response (Figure 5.2).

Positive reinforcement is the scenario in which the performance of a specific behavior is reinforced by adding something pleasant, so that the overall consequence of that particular behavior is an improvement of the animal's environment or mental state. Such a consequence will reinforce the behavior that elicited the improvement, making it more likely that the animal will repeat it, should the chance arise. Negative reinforcement is a scenario in which a particular behavior is reinforced, because something is removed/subtracted ("negative") from the animal's environment as a consequence of performing the behavior. For this to happen, the stimulus that the animal can make disappear by showing a particular behavior, must be something unpleasant. The stronger the reinforcing properties of the consequence of performing the behavior the faster and stronger the conditioning will be. This characteristic is different from the case of the salience of the CS, where it was only the speed and not the strength of the conditioning that was affected.

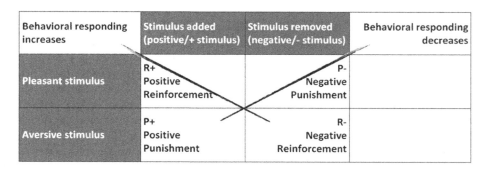

Behavioral responding increases	Stimulus added (positive/+ stimulus)	Stimulus removed (negative/- stimulus)	Behavioral responding decreases
Pleasant stimulus	R+ Positive Reinforcement	P- Negative Punishment	
Aversive stimulus	P+ Positive Punishment	R- Negative Reinforcement	

FIGURE 5.2 The four operant procedures. Positive reinforcement and negative reinforcement will lead to an increase in the probability of the behavior recurring, whereas positive punishment and negative punishment will lead to a decrease in the probability of the behavior recurring.

Positive punishment and negative punishment are both scenarios in which the likelihood of the animal repeating the eliciting behavior will decrease; the behavior is weakened. As with reinforcement, punishment can happen by either the addition of something unpleasant (positive punishment) or the removal of something pleasant (negative punishment).

The concept of escape/avoidance is sometimes used to denote two of these operant procedures functioning by manipulating aversives. When an aversive stimulus (e.g., an electric shock following a tone in compartment A in a two-compartment shuttle box), the rat will – on repeated trials – learn that when the tone sounds, she/he must quickly run to the other compartment (compartment B) to avoid the shock. In that case, the rat now avoids the aversive stimulus (the shock). The rat has learned to predict the aversive stimulus by associating it with the CS (the preceding tone). By showing the appropriate behavioral responses to the stimuli predicting the unpleasant event, the rat avoids the shock.

The same applies to the response of a cat sharpening her/his claws on a leather couch. When the owner squirts the cat with a squirt gun, the cat runs away. In time, the cat will learn to keep her/his claws off the furniture, when the discriminative stimulus, the owner, and/or the squirt gun is present.

These are examples of positive punishment in which the animal escapes the aversive stimulus. The clawing on the couch and the staying in compartment A are both behaviors that are positively punished and therefore the likelihood of the animal showing the same behavior again is reduced. At the same time, the cat's "not-clawing-the couch-and-running-away" behavior is negatively reinforced as the aversive stimulus is removed when the cat shows that behavior. The rat jumping into safety in compartment B is also negatively reinforced. Both animals have learned to predict the aversive stimulus (water squirt or shock) by associating it with the CS (squirt gun or tone). By showing the appropriate behavioral responses to the stimuli predicting unpleasant events, the cat avoids the water squirt and the rat avoids the shock.

When training animals using operant conditioning, it is not possible to work only with one of the four operant procedures. In the above examples, positive punishment and negative reinforcement are both affecting the behavior of the animals. If positive reinforcement, on the other hand, is the primary procedure, any correct responses will be positively reinforced; however, any incorrect responses will result in the trainer withholding/taking away the reinforcer, which is a negative punishment of the incorrect behavioral response. It is of course also possible to combine all four operant principles.

Moreover, the above examples also demonstrate that it is not possible to separate classical conditioning and operant conditioning when training animals (please also refer to Chapter 6).

It is important to note that the perceived valence (good or bad) of a stimulus may differ between species and may vary from individual to individual. The easiest way to assess the animal's perception of the operant consequence is to see whether the behavior is strengthened or weakened. For example, imagine a researcher entering a pigpen housing three-month-old slaughter pigs and one pig bites the researcher's boots. The researcher wants that behavior to stop and the researcher starts pushing the pig away while yelling "bad pig." He is trying to punish the unwanted behavior, as pushing and yelling are considered unpleasant stimuli (by the researcher). The pig, however, enjoys the whole hullabaloo enormously; it is the most fun that has happened in several days – it may even be like playing with other pigs – and hence the boot-biting behavior is reinforced. The more the researcher pushes the pig to make her stop, the more boot-biting the pig will do. Obviously, it is the pig that decides whether a consequence is a reinforcer (which will then result in more behavior) or a punisher. It is not something the researcher decides.

In operant conditioning, the animal hence learns an association between her behavior and the resulting consequence. However, it is far from always that a specific behavior will lead to reinforcement and the animal will quickly learn to associate certain stimuli – discriminative stimuli, S^ds – in the surroundings with whether or not the behavior will be reinforced. It is interesting that some stimuli seem to be better S^ds than other stimuli. In 1973, Foree and LoLordo trained pigeons to press a treadle (a lever device pressed by the foot) using a combination of a tone and a red light as discriminative stimuli.[14] When the sound and light was on, pressing the treadle would result in access to grain in one group of pigeons. In another group of pigeons, pressing the treadle would result in the non-appearance of an electric shock. When the pigeons had learned this task, the two stimuli (the tone and the light) were tested separately, and it turned out that the tone and the light did not have the same effect on the subsequent behavior. In pigeons working for grains, the birds would press the treadle more when the light was used as S^d. The tone alone only resulted in few responses. If the pigeons were pressing the treadle to avoid getting shocked, the pigeons would respond more to the tone than to the light; i.e., for an avoidance-task the pigeons seemed to have focused on the sound, whereas for the appetitive, food-related task, the birds had primarily used the light as their cue. This is a similar, although less drastic, example of the same principles that were demonstrated in the taste aversion conditioning experiment mentioned in the section on classical conditioning.

EXTINCTION IN OPERANT CONDITIONING

If a learned behavior is no longer reinforced it will disappear over time. If a cat has learned to scratch the door to get a human to open it, the scratching behavior will eventually stop if the human totally ignores the scratching and does not open the door anymore when the cat scratches it. The cat will now learn something new: that scratching the door does not have any consequences. However, an important feature of extinction is the extinction burst. If the reinforcement expected by the animal is no longer a consequence of the behavior, the animal will first intensify the behavior. If the human suddenly does not open the door when the cat scratches the door, the scratching will increase in intensity and/or frequency and even other behaviors such as vocalization may appear. Thereafter, if nothing happens, the behavior will stop. The key point here is that if the human gives in during the extinction burst and opens the door, the cat will learn to intensify the behavior to be reinforced and extinction may become more difficult. This strategy is often used in animal training, when the trainer wants the animal to increase intensity or frequency of a behavior. By withholding the reinforcement, the trainer will prompt the animal to show more behavior, which is then reinforced. When an animal trainer is using this technique, it is vital to minimize frustration and hence the trainer's timing is pivotal (please refer to Chapter 6 for more details).

SCHEDULES OF REINFORCEMENT

Schedules of reinforcement are rules that decide the relationship between the behavioral response and the reinforcer. If a reinforcer is delivered every time a specific behavioral response is performed, it is a continuous schedule of reinforcement (CRF). However, a specific behavioral response is not necessarily reinforced every time, but on an intermittent (or partial) schedule. Sometimes it is necessary to show a specific response several times before the reinforcer is delivered – and sometimes reinforcement will only be available after a certain period has elapsed. The type of contingency (i.e., whether a behavior is reinforced on a continuous or an intermittent schedule of reinforcement) will influence the animal's behavior. Basically, there are two types of intermittent schedules; either the relationship is based on number of responses needed to obtain the reinforcer (i.e., the ratio between number of responses and reinforcer) or the time elapsed since the last reinforcer was available (i.e., an interval). The schedules may also be either fixed or variable. A Fixed Ratio (FR) schedule is a reinforcement design, where a pre-set (fixed) number of responses are required to obtain the reinforcer. For example, a rat must press the lever in the Skinner box precisely 20 times to obtain the food pellet and thus the ratio is 20 responses to one reinforcer. This is a fixed ratio of 20/1 or simply FR20. If the response must be shown three times before delivery of the reinforcer, it is an FR3 and if every response is followed by a reinforcer, it is an FR1 and equivalent to a continuous schedule of reinforcement (CRF). If, on the other hand, the reinforcer is delivered after an average number of responses have occurred, the schedule of reinforcement is a variable-ratio schedule. If a rat – in order to obtain a reinforcer – must press the lever 17 times, then 23 times, 16 times, and then 24 times, then the average responses are 20 presses per reinforcer. This schedule is called a variable ratio 20 or simply VR20. Compared to fixed ratio schedules, variable ratio schedules maintain high and steady rates of the desired behavior, and the behavior is more resistant to extinction.[15,16]

Fixed Interval schedules involve reinforcing a specific behavior after a fixed interval of time has elapsed. Hence, on a Fixed Interval (FI) three-second schedule, the first response after three seconds have passed from the presentation of the cue will be reinforced. Any responses made before the three seconds have passed will not be reinforced. In a variable interval schedule (VI), the interval of time is not always the same, but the average length of the time-interval will define the schedule. For example, in VI8, the required time interval may be four seconds on the first trial, and then 12, 10, 6, and 8 … and thus the average time interval will be eight. In animal training in zoos, in laboratory facilities, and pet animals, ratio schedules are the most used schedules of reinforcement and especially in zoos, aquariums, and laboratories, CRF should be the preferred reinforcement schedule (see also Chapter 6).

EFFECT OF POSITIVE REINFORCEMENT IN TRAINING

Research done primarily in apes and non-human primates has shown that training the animals using primarily positive reinforcement can result in animals that willingly and voluntarily participate in procedures that are normally considered unpleasant for the animal.[17,18] Moreover, trained animals show less stress- and fear-related behavior when participating in experimental procedures.[19,20] Positive Reinforcement Training (PRT) can also lead to improved research results[21,22] and might reduce the need for restraint or anesthesia.[23,24] In some cases, the drug dose needed for anesthesia may also be reduced; most likely because the animal is less stressed (C. Grøndahl, chief veterinarian in Copenhagen Zoo, personal communication). Moreover, in a study by Leidinger et al. (2007) it was proposed that PRT has the potential to reduce the variation of behavioral responses in mice, and thus reduces the number of animals needed to attain valid test results.[25] Lastly, studies that otherwise could not have been done (e.g., catheter blood sampling and catheter maintenance on a calm, un-sedated pig standing still for the procedures or an un-sedated pig standing still for multiple x-rays on a platform) has been completed after training the pigs to cooperate.

EFFECT OF POSITIVE PUNISHMENT AND NEGATIVE REINFORCEMENT IN TRAINING

Positive punishment (P+) works by evoking irritation, discomfort, or even fear or pain in the animal and the response of the animal may sometimes be species-specific defense reactions such as freezing, fleeing, or aggression. In negative reinforcement training, aversive stimuli are also applied (initially often at increasing intensity to create the escape response), but the animal learns to avoid the unpleasantness. The use of aversive stimuli in P+/R− paradigm may risk inducing a negative emotional response in the animal and thereby compromise the training.

Research has shown that negative reinforcement training and positive punishment training can result in escape-related behaviors, aggression, and fear reactions in dogs[26] and this may very well be applicable for other animal species as well. These effects of the use of aversives in animal training should not be surprising, as all living beings will try to avoid unpleasant stimuli. The stronger the negative event the greater the risk that it will create problems. Using low levels of aversive events in combination with letting the animal control the situation will however not necessarily result in either stress in the animal or negative responses to the trainer.[27] It is worth noting that experience from the zoo and aquarium world emphasizes the importance of not inducing frustration, fear, or aggression in the animals during training, as it may severely compromise the safety of the trainer. These risks are of course dependent on the size, strength, and temperament of the animal but also on domestication. However, just because an animal does not impose an immediate risk to the trainer or handler, the welfare of the animal should still be highly prioritized.

When working with laboratory animals the aim should always be to enhance welfare as far as possible. Positive reinforcement training using classical and operant conditioning techniques can thus be used to train animals to cooperate, making the procedures both predictable, controllable, and rewarding for the animals. There is really no valid excuse for not implementing animal training in some form in all animal facilities. However, this demands skilled trainers with extensive knowledge of both scientific learning principles and the practical application of these principles.

REFERENCES

1. Groves, P. M. & Thompson, R. F. Habituation: A dual process theory. *Psychological Review* **77**, 419–450 (1970). doi: 10.1037/h0029810
2. Obernier, J. A. & Baldwin, R. L. Establishing an appropriate period of acclimatization following transportation of laboratory animals. *Ilar Journal* **47**, 364–369 (2006). doi: 10.1093/ilar.47.4.364
3. Graham, J. M. & Desjardins, C. Classical conditioning: Induction of luteinizing hormone and testosterone secretion in anticipation of sexual activity. *Science* **210**, 1039–1041 (1980). doi: 10.1126/science.7434016
4. Wirth, T. et al. Repeated recall of learned immunosuppression: Evidence from rats and men. *Brain Behavior and Immunity* **25**, 1444–1451 (2011). doi: 10.1016/j.bbi.2011.05.011
5. Estes, W. K. & Skinner, B. F. Some quantitative properties of anxiety. *Journal of Experimental Psychology* **29**, 390–400 (1941). doi: 10.1037/h0062283
6. Garcia, J. & Koelling, R. A. J. P. S. Relation of cue to consequence in avoidance learning. *Animal Behavior* **4**, 123–124 (1966). doi: 10.3758/bf03342209
7. Coombes, S., Revusky, S. & Lett, B. T. Long-delay taste aversion learning in an unpoisoned rat: exposure to a poisoned rat as the unconditioned Stimulus. *Learning and Motivation* **11**, 256–266 (1980). doi: 10.1016/0023-9690(80)90016-8
8. Garcia, J., Ervin, F. R. & Koelling, R. A. Learning with prolonged delay of reinforcement. *Psychonomic Science* **5**, 121–122 (1966). doi: 10.3758/BF03328311
9. Garcia, J., Kimeldorf, D. J. & Koelling, R. A. Conditioned aversion to saccharin resulting from exposure to gamma radiation. *Science* **122**, 157–158 (1955). doi: 10.1126/science.122.3160.157
10. Rescorla, R. A. Conditioned inhibition of fear resulting from negative CS-US contingencies. *Journal of Comparative and Physiological Psychology* **67**, 504 (1969). doi: 10.1037/h0027313
11. Jordan, W. P., Todd, T. P., Bucci, D. J. & Leaton, R. N. Habituation, latent inhibition, and extinction. *Learning & Behavior* **43**, 143–152 (2015). doi: 10.3758/s13420-015-0168-z
12. Gamzu, E. & Williams, D. R. Classical conditioning of a complex skeletal response. *Science* **171**, 923 (1971). doi: 10.1126/science.171.3974.923
13. Thorndike, E. L. *Animal Intelligence: Eksperimental Studies* (New York: The Macmillan Company, 1911).
14. Foree, D. D. & Lolordo, V. M. Attention in the pigeon: Differential effects of food-getting versus shock-avoidance procedures. *Journal of Comparative and Physiological Psychology* **85**, 551–558 (1973). doi: 10.1037/h0035300
15. Domjan, M. In *The Principles of Learning and Behavior* Ch. 9 (Belmont, CA: Wadsworth, Cengage Learning, 2010).
16. Crossman, E. K., Bonem, E. J. & Phelps, B. J. A comparison of response patterns on Fixed-ratio, variable ratio, and Random-ratio schedules. *Journal of the Experimental Analysis of Behavior* **48**, 395–406 (1987). doi: 10.1901/jeab.1987.48-395
17. Freymond, S. B. et al. Behaviour of horses in a judgment bias test associated with positive or negative reinforcement. *Applied Animal Behaviour Science* **158**, 34–45 (2014). doi: 10.1016/j.applanim.2014.06.006
18. Innes, L. & McBride, S. Negative versus positive reinforcement: An evaluation of training strategies for rehabilitated horses. *Applied Animal Behaviour Science* **112**, 357–368 (2008). doi: 10.1016/j.applanim.2007.08.011
19. Westlund, K. Training is enrichment-And beyond. *Applied Animal Behaviour Science* **152**, 1–6 (2014). doi: 10.1016/j.applanim.2013.12.009
20. Perlman, J. E. et al. Implementing positive reinforcement animal training programs at primate laboratories. *Applied Animal Behaviour Science* **137**, 114–126 (2012). doi: 10.1016/j.applanim.2011.11.003
21. Schapiro, S. J., Perlman, J. E., Thiele, E. & Lambeth, S. Training nonhuman primates to perform behaviors useful in biomedical research. *Lab Animal* **34**, 37–42 (2005).

22. Kemp, C. et al. A protocol for training group-housed rhesus macaques (Macaca mulatta) to cooperate with husbandry and research procedures using positive reinforcement. *Applied Animal Behaviour Science* **197**, 90–100 (2017). doi: 10.1016/j.applanim.2017.08.006

23. Gillis, T. E., Janes, A. C. & Kaufman, M. J. Positive reinforcement training in squirrel monkeys using clicker training. *American Journal of Primatology* **74**, 712–720 (2012). doi: 10.1002/ajp.22015

24. Coleman, K. et al. Training rhesus macaques for venipuncture using positive reinforcement techniques: A comparison with chimpanzees. *Journal of the American Association for Laboratory Animal* **47**, 37–41 (2008).

25. Leidinger, C., Herrmann, F., Thoene-Reineke, C., Baumgart, N. & Baumgart, J. Introducing Clicker Training as a Cognitive Enrichment for Laboratory Mice. *Jove-Journal of Visualized Experiments* (2017). doi: 10.3791/55415

26. Ziv, G. The effects of using aversive training methods in dogs: A review. *Journal of Veterinary Behavior: Clinical Applications and Research* **19**, 50–60 (2017). doi: 10.1016/j.jveb.2017.02.004

27. Wergard, E. M. et al. Training pair-housed Rhesus macaques (Macaca mulatta) using a combination of negative and positive reinforcement. *Behavioural Processes* **113**, 51–59 (2015). doi: 10.1016/j.beproc.2014.12.008

6 Animal Training
The Practical Approach

*Dorte Bratbo Sørensen, Annette Pedersen,
and Robert E. (Bob) Bailey*

CONTENTS

INTRODUCTION

Animals are constantly learning, and animal training is a way of guiding that learning in a direction that is beneficial for our ability to care for and create relationships with the animals. Animal training as a concept is often used broadly and with different meanings. Sometimes it may refer to habituation or socialization and sometimes to successful ways of handling animals. In this chapter animal training is defined as the modification of an animal's behavior by the systematic and well-planned use of operant and classical conditioning techniques.

The practical approach to animal training should be science-based. The training methods should build on scientific studies demonstrating the effect of stimuli on the behavior of animals both in associative and non-associative learning (see Chapter 5). Among the pioneers, on whose work the training techniques are built, were B.F. Skinner, E.L. Thorndike, and Ivan Pavlov, who – by modifying the animals' environment – were able to change their behavior through careful, hypothesis-driven research. These pioneers established the principles of operant and classical learning and were able to predict and modify the behavior of animals.[1–3] These scientific principles form the basis of various practical disciplines such as behavior analysis in human psychology and animal training technology and methodology.

To ensure future professional development of science-based practical animal training, animal trainers worldwide should use the same terminology to communicate, teach, and discuss all the practical elements of animal training. A consistent terminology is important to promote clarity and avoid misunderstandings. For example, the word "rewarded" is often used instead of or interchangeable with the word "reinforced" and in the same manner the "reinforcer" is sometimes referred to as a "reward" and vice-versa. However, often the word "reward" is also used in a broader sense, meaning "something that is given in return for good or evil done or received or that is offered or given for some service or attainment" (as defined by Merriam-Webster online dictionary). A reward is thus not necessarily given to increase the possibility of a certain behavior. A reward may of course in some situations have a reinforcing effect, whereas at other times it is merely a consolation. When discussing stimuli specifically linked to an operant procedure in a training set-up, we suggest always using the term "reinforcer." In the same way, the word "punishment" may by some readers be considered to indicate a severe, aversive event, perhaps even something that may injure the punished individual. The terms "punishment" and "punisher" as used in this chapter relate solely to the effect on operant behavior. In other words, a punisher is merely a stimulus that, when applied as a consequence of a particular behavior, reduces the possibility of that behavior being repeated.

The science behind animal training entails both classical and operant conditioning (see Chapter 5), but trainers often tend to focus on the operant aspects such as cues, criteria, and rate of reinforcement. Hence, the trainer's awareness will often be drawn towards the target behavior (the behavior the trainer is aiming to get) and how to reinforce it. However, the biological principles of classical conditioning must not be underestimated as they are constantly in play when animals learn about their environment. Depending on the animal's previous experience and learning, emotional responses may be elicited by stimuli in the environment. If, for example, a dog has – as a puppy – been beaten with a broomstick, presenting a target stick to the dog may elicit a fear reaction. Such emotional responses may sometimes be powerful enough to override the operant response the trainer is working on conditioning. The Bailey-Breland proverb "Pavlov is always sitting on your shoulder" is forever valid – it is not possible to train an animal using solely operant conditioning; ignoring the principles of classical conditioning is detrimental. Even though the trainer should know the science behind the techniques, several practical details and considerations need to be addressed in order to be successful. In this chapter we will present a selection of these considerations building on decades of experience training animals as well as animal trainers.

THE ABC OF ANIMAL TRAINING

One way to address animal training is by the use of the ABC model – the ABC of behavior. The origin of this model fades into the mists of time, but it seems to have sprung from the field of behavioral analysis in the 1980s. The ABC of behavior encapsulates the essence of animal training perfectly, as A stands for Antecedents (all that comes before the behavior), B stands for the target Behavior, and C stands for the Consequences (both good and bad) of that behavior.

The Antecedents include everything that affects the animal or has affected the animal previously. This includes the genetics of the animal, all previous experiences (including training and other experiences with humans), health, mental condition, and the environment in which the animal is situated (both physical and social). If a behavioral response has already been trained, the antecedent also includes the discriminative stimulus (denoted Sd or sometimes S+) indicating that a

FIGURE 6.1 The Skinner box (an operant conditioning chamber) may be designed in a large variety of ways. This Skinner box is a simple, traditional box equipped with a house light (A), a response lever (B), a stimulus light (C), and a delivery device for reinforcers – in this case a water nipple (D). Both A and C may function as discriminative stimuli. For example, when the house light is on, the system is activated, and reinforcement may be available with certain intervals. When the stimulus light (C) is on, pressing the lever will result in delivery of a reinforcer. When the light is off, no reinforcement will be delivered for lever pressing.

certain behavior will have reinforcing consequences. Such discriminative stimuli are also known as signals or cues. For example, if pressing a lever in a Skinner box (Figure 6.1) when the stimulus light is on produces food, then the lit stimulus light is an Sd. Another stimulus (denoted SΔ, s-delta or S−) may signal that a response will not be reinforced. For example, if the stimulus light over the lever is off, the lever pressing will not be reinforced; hence the unlit stimulus light is a SΔ. Both are discriminative stimuli, preceding the behavior and at the same time conveying information about the consequences of a specific behavior.

The B (the Behavior) denotes the desired behavioral response, the target behavior. For any training session it is highly important to thoroughly define the behavioral response that will be reinforced (the criteria, please refer to the section on Criteria). If the trainer is not absolutely confident what she/he will and will not reinforce, the trainer may risk being a little hesitant, which in turn will affect the timing, speed, and accuracy of the training.

The C (the Consequence) is the stimulus/stimuli elicited by the display of the behavioral response. In a training situation, the consequence of performing a behavior requested by the trainer is often (but not always) access to a preferred food item. It is important to realize that the Consequence of a certain behavior subsequently adds to the Antecedents. This means, for example, that if a trainer has employed an aversive positive punishment as a Consequence of the animal showing an unwanted behavior, the animal will add the unpleasantness of the punisher in that particular situation or context to his experiences and bring it into the next training session. Conversely, if the animals' behavior has been reinforced with a valuable reinforcer, the pleasantness of that particular training session will positively affect the animals' perception of the next training session.

THE CONSEQUENCES OF BEHAVIOR

There are two possible consequences of showing a behavior, either something good or something bad happens. It is important to be aware that the value of this consequence will affect any expectations that the animal may have. If a monkey expects to obtain a grape, a large piece of cucumber is bad, whereas if she expects a small piece of cucumber, a large piece of cucumber is good. It is imperative to remember that it is not the trainer who decides whether a consequence is good or bad; it is the animal. Some consequences may be reinforcing to some individuals (e.g., getting pizza when you helped a friend move), but have no value for others (getting pizza when you helped a friend move and your friend forgot that you are lactose and gluten intolerant).

THE REINFORCERS

A primary reinforcer is anything an animal will approach and/or work for. Whether it is a stimulus the animal needs, wants, or likes is less relevant for the purpose of this chapter; the key point is that the animal is motivated to perform a behavioral response to obtain the stimulus. Some primary reinforcers are innate (like food, water, heat when you are cold, a safe nesting site, etc.), whereas others may be stimuli that at first were not necessarily pleasurable for the animal, but over time have become something the animal appreciates. For example, in some species, such as white rhinos, all individuals like to scratch themselves; however, if scratching is to function as a primary reinforcer provided by the trainer in the training sessions, it requires a good relationship between the rhino and the trainer. In dogs, being reinforced with a toy that facilitates play behavior seems to work instantly in some animals whereas others (of the same breed, age, and gender) may take a little more time to realize the full, enjoyable potential of playing with that toy. Moreover, reinforcers must meet the animals' expectations. The animal quickly learns what to expect and if the delivered reinforcer has too low a value (compared to what the animal was expecting), the level of motivation will decrease, and the animal may stop responding. If the behavior is reinforced above the level of expectation, the level of motivation increases. It is thus important that the trainer is able to judge the animal's level of motivation to work for the reinforcer and to make any adjustments needed. Some trainers distinguish between innate primary reinforcers and reinforcers that the animal, over time, has learned to value. However, it seems superfluous to split up the trainer's toolbox of primary reinforcers. It all comes down to knowing what the individual animal will and will not work for. The nature of the primary reinforcers may thus vary due to personality traits, experience, and preferences of the individual animal.

A conditioned reinforcer (also known as a secondary reinforcer) is a reinforcer which initially has no value. Through classical conditioning this stimulus has been associated with a primary reinforcer and hence gained reinforcement value (also see Chapter 5). Some conditioned reinforcers (e.g., the verbal "good boy") may over time and with experience gain enough strength to become a primary reinforcer. Whether this will happen depends on the stimulus, the individual, and the relationship between the animal and the trainer.

A conditioned reinforcer such as a clicker (a small device that makes a distinct, short click sound) or a whistle can be used to accurately mark the specific behavior that the trainer is aiming for. Such a marker can be useful, e.g., when training an animal to place both paws on an elevated station. The use of the marker will allow the trainer to precisely reinforce the intention to lift a paw (i.e., to click when the animal shifts the weight from one leg to another, prior to the paw lifting) and gradually ask for more and more until the animal actually lifts the paw off the ground. The conditioned reinforcer is also – especially by zoo and aquarium trainers[4] – called a "bridge;" a stimulus that bridges the behavior of the animal and the primary reinforcer. Using a conditioned reinforcer requires it to be short and salient, such as a clicker or a short sound of a whistle. It is not mandatory to use such devices; however, depending on the animal, the behavior, and the environment, a conditioned reinforcer may significantly increase the chance of training success.

When starting the training, it is important to ascertain that the conditioned reinforcer is in fact conditioned, e.g., that the animal has associated the clicker with the food. However, before conditioning the clicker it is vital to have a continuous and fluent consumption of the primary reinforcer. In other words, the animal must demonstrate that the primary reinforcer is in fact preferred and something she/he will accept without caution or reluctance.

Any stimulus that predicts delivery of the primary reinforcer can become a conditioned reinforcer. Especially in inexperienced trainers, small body movements may precede the click. For example, imagine that a trainer's hand is reaching for the food bag just prior to the click as the animal gives a response. Even though the trainer may not be aware of this body movement, the animal may see it and hence it can serve as a conditioned reinforcer. Such inadvertent conditioned reinforcers should be avoided as it will jeopardize the timing in the training since the trainer is now not consciously controlling the conditioned reinforcers. It does not matter whether the trainer's hand movement is conscious or not. What matters is that the hand movement can be a visual stimulus to the animal predicting that food is on the way. The animal will therefore be more attentive to body movements and will, e.g., have difficulties working if she/he cannot see the trainer.

A conditioned reinforcer functions as a signal predicting the primary reinforcer and will evoke a state of positive anticipation.[4] If this anticipation is not met, a quick decrease in the function of the conditioned reinforcer as a predictor signal will be seen. In other words, if the association formed through classical conditioning is not maintained, extinction will take place. Hence, the association between the primary and the conditioned reinforcer needs to be upheld (a simple rule-of-thumb is: "when you click, you feed"), so that every presentation of the conditioned reinforcer is followed by a primary reinforcer.

When the primary as well as the conditioned reinforcer has been decided, the trainer should carefully consider how to deliver the primary reinforcer. This is highly dependent on the animal species, the environment, and the reinforcer itself. A prime principle is that the reinforcer must be delivered fast and accurately. Moreover, it is often preferable that the time it takes for the animal to consume the reinforcer is as short as possible. As much thought should be given on where to deliver the primary reinforcer as to when and how. The reinforcer should always be delivered to the animal in such a way that the animal is positioned optimally for the next response (as expressed by the Bailey-Breland proverb "click for action, feed for position."). For example, if the trainer wants an animal to move forward, she/he clicks the movement forward and delivers the primary reinforcer in front of the animal. If the trainer wants the animal to back away, the movement backwards is clicked, and the primary reinforcer is delivered between and slightly behind the animal's front leg to keep the animal in a "backwards-moving" state. In this way, the animal is "clicked" for moving back and fed in a way that keeps the animal in a position ready to move further back.

THE AVERSIVES

Primary aversives (or punishers) are stimuli that the animal will escape or avoid. It is imperative to note that in the context of operant learning, an aversive is not necessarily a stimulus that will harm the animal; a stimulus being mildly unpleasant or just annoying such as a light pressure may be sufficient to modify behavior. Moreover, it must be noted that if an animal in a training setup has the choice between two consequences (e.g., a foot shock if not responding and being abruptly lifted back into a start box by the trainer if showing the correct response), then both these consequences are aversive to the animal; however the lifting is less aversive than the shock and hence will function in the situation as a reinforcer. Just as with reinforcers, aversives can either be primary or conditioned.

THE IMPORTANCE OF TIMING, CRITERIA, AND RATE OF REINFORCEMENT

TIMING

Timing is the trainers' ability to present the reinforcers (primary and conditioned) at the precise moment when the animal shows the correct response. For a trainer to do this, the trainer must be able to anticipate behavior, which requires a thorough knowledge of the species and individual at hand. Animals are efficient problem solvers and most species will learn a specific behavioral response in spite of a trainer with bad timing. The caveat is that the bad timing will frustrate the animal, delay the training, and compromise the outcome of the training. Two factors will determine which behaviors are reinforced: the ability of the trainer to time a conditioned reinforcer and the speed with which the trainer can deliver the primary reinforcer. Consider a chicken learning to peck a target. If the trainer's timing is bad, she/he may either click too early (when the chicken's beak is going towards the target but not yet touching) or too late (when the chicken is moving her head upwards after the peck). In the first scenario the resulting behavior after a few sessions will be a chicken pecking in the air right above the target and never actually touching the target. In the latter scenario, the chicken lowers her head and then abruptly lift it as it is the head lift that has been reinforced. The speed with which the trainer can deliver the primary reinforcer is also vital. If the trainer is too slow the animal may show other behaviors in the time span between the conditioned reinforcer (e.g., the click) and the primary reinforcer (e.g., a piece of apple). Consider, for example, a pig trained to follow a target. The trainer plans to click for the pig following a moving target for two meters. The trainer clicks, but while reaching for the apples, the pig bites the target, lets go again and starts rooting ... and then she gets the primary reinforcer. Then the trainer will actually have reinforced both "biting the target" and "rooting" as well as "follow target" (Figure 6.2).

CRITERIA

A criterion is the behavior that the trainer will reinforce in a particular session and all criteria should be clearly defined (e.g., jump 10 cm higher; peck harder so the ball moves; hold position for 12 seconds, etc.). An important aspect of setting the criteria is the observational skills of the trainer. If the trainer has poor observational skills, the trainer will risk being inconsistent in the reinforcement of the desired behavior. Moreover, many trainers are emotionally involved while training, which may risk influencing accepting or not accepting a given behavior.

In general, the criterion should be set so the trainer expects the animal to achieve 80–90% success. When the animal is responding correctly eight times out of ten, the criterion can be raised, making the task a little more difficult. The trainer should consistently work on only one criterion at the time and not ask for, e.g., both standing still for a longer time (criterion 1) and accept poking with a pen for the first time (criterion 2) in the same session. Some trainers may accept as low as 40–60% correct responding for very simple behaviors; however, it requires a very high level of trainer experience and skills to do so without inducing adverse effects such as frustration or aggression in both trainer and animal. In general, it is not advisable to work with such low success rates. A simple way to keep track on the success rate is to bring ten reinforcers for the training session. If all ten responses are correct, then all ten reinforcers will be used when the session ends. If three reinforcers still remain after ten cues have been given, then the animal has responded incorrectly three times and the success rate is 7/10 = 70%.

RATE OF REINFORCEMENT

The Rate of Reinforcement (RoR) denotes the number of reinforcers given per time unit (e.g., ten reinforcers per ten seconds compared to two reinforcers per ten seconds). The rate of reinforcement depends on a large number of factors; however, there will – under a specific set of circumstances

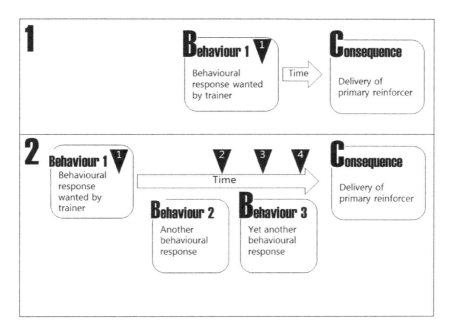

FIGURE 6.2 (1) The conditioned reinforcer (e.g., a clicker-sound, marked with black triangle-arrow) must be timed precisely when the correct response is shown (triangle-arrow 1). (2) If the trainer is too slow, timing of the conditioned reinforcer may be wrong and behaviors other than what the trainer planned will be reinforced. If for example the trainer clicks at arrow 2, the trainer will reinforce behavior 2 (conditioned reinforcer) and behavior 3 (primary reinforcer). If the trainer clicks at arrow 4, behavior 3 will be reinforced both by the conditioned reinforcer and the primary reinforcer. The primary reinforcer, which most likely is the strongest, may reinforce behavior 3 the most and may reinforce the actual trained response (behavior 1) the least. It is important to remember that in a scenario like this, all behaviors (behavior 1, 2, and 3) are more or less reinforced; it is not just the behavior immediately prior to the reinforcer. In other words: if the animal has time to perform other behaviors in the gap between the conditioned reinforcer and the delivery of the primary reinforcer, these behaviors will also be reinforced … and possibly more than the behavior the trainer thinks she/he is reinforcing. The above examples may be modified by previous reinforcement history and the nature of the reinforced behavior.

– be an optimal rate of reinforcement. If the RoR drops and becomes very low, this may result in decreased motivation (maybe causing the animal to leave the trainer) or frustration that may result in unwanted behavior and even aggression. The RoR may also drop if the animal is no longer motivated, e.g., if the animal is satiated and no longer interested in the primary reinforcer (if it is food). The skilled trainer will make sure to end the training session prior to the animal losing interest. If the RoR becomes very high, it could be that the criterion is set too low and the task is very easy for the animal to perform.

Whenever the trainer enters a training session it must be absolutely clear to the trainer which behaviors will be reinforced and which behaviors will not. If the trainer has not decided beforehand, she/he will need to mentally consider each behavior offered by the animal, which will negatively affect the timing. If the training does not proceed as planned, the antecedents and the criteria set for the session should be re-evaluated. Trainers would benefit significantly by video recording training sessions and reviewing them afterwards and/or having another trainer watch training sessions, looking for trainer errors. Moreover, a thorough and well-considered plan for the training should be constructed. Such a plan could include, e.g., the final behavior, the intermediate steps towards the final behavior, chosen reinforcers (primary and conditioned reinforcer, if used), the criteria for each stepwise behavioral response trained, and how to chain the trained behaviors, if needed (Figure 6.3, A).

PLAN FOR TRAINING Name of trainer		Page
Animal ID/Name		
Species/breed/strain	Sex	Age
Goal behavior (in details)		
Step-by-step plan for getting the behavior (capturing, shaping, luring) incl. plan for fading of prompts or lures.		
Training sessions and criteria (Training session: how often, duration of session; define a criterion for each step).		
Reinforcer	Bridge	
Trainer		

TRAINING LOG Name of trainer			Page
Animal ID/Name			
Species/breed/strain	Sex		Age
Goal behavior			
Reinforcer		Bridge	
Date	Criteria, success rate, details on animal, other relevant		Trainer Initials

A   **B**

FIGURE 6.3   A) Training plan and B) Log for training.

GETTING THE BEHAVIOR

When working with an animal to modify behavior, the obvious and most important thing is to get the animal to show the behavior the trainer wants. If the trainer fails this mission, e.g., by setting the criteria too high, the rate of reinforcement will be very low, and the animal may lose interest or react with frustration or even aggression. Therefore, the ability of the trainer to get the animal to show the correct behavior is an important skill. Arranging the antecedents (e.g., manipulate the environment or criteria) to make it as easy as possible for the animal to succeed, or, said another way, make it difficult for the animal to err is vital. For example, if the animal is kept in a large cage and the trainer wants him positioned along the bars in the front of the cage, obstacles can be placed to physically guide the animal to the correct position (Figure 6.4).

FIGURE 6.4   A) The positioning of a big log in the cage to bring the bear close to the bars. B) Bear being correctly positioned close to the trainer and the bars due to the careful placing of the big log.

When the animal has learned the correct position, the obstacles can be gradually removed. It is also possible to make use of an animal's natural tendencies and preferences to get the behavior. For example, if a cat needs to be weighed, move the scale from the ground and place it on an elevated platform as jumping up and sitting high is often preferred behavior for a cat.

Depending on the animal, the training history, and the relevant antecedents, several strategies are available for getting the behavior, namely capturing, targeting, luring/baiting, and shaping. Currently, in the laboratory facility, targeting seems to be the most used technique to get the behavior, though luring/bating and shaping is probably also used.

CAPTURING

Capturing is used if the animal will present the full behavior in one trial without being lured to do the behavior. This will often be easy with simple behaviors such as a dog sitting, whereas behaviors such as yawning or sneezing may require a lot more patience from the trainer. The key point in capturing is that the trainer reinforces only the full behavioral response and a trainer with experience and good observational skills can sometimes capture all, or nearly all, of the behavior with a few trials. Moreover, no intermediate behavioral responses are reinforced and therefore added to the animals' behavioral repertoire. Any intermediate responses that have been reinforced during the course of the training may have to be extinguished later on. However, using capturing that will not be needed.

In some instances, trainers will provoke a behavioral response in order to capture it. Such a technique requires a lot of experience, since the provocation of a behavior in the beginning may induce what could be called annoyance in the animal; a preference for creating distance from the trainer, either by the animal actively moving away or by the animal forcing the trainer to move away. When seals are trained to wave their front flipper, the trainer builds on a behavior which is in fact a precursor of aggression in seals, namely the front flipper waving. By withholding food or otherwise annoying the animal, the trainer provokes the wave, which should be precisely reinforced. During the training, the initial negative emotional response connected to the behavior is modified, and the seal's aggression decreases and eventually disappears even though the behavior per se stays intact. Using capturing should always entail preceding considerations on the possible negative emotional effect – if any – on the animal.

TARGETING

Targeting (also known as target training) is the technique of teaching the animal to go to, and maybe touch, a specific object in order to obtain reinforcement. The animal learns to "target" on that object, and the object can be moved around with the animal following. The target may also be used to position the animals. For instance, large boars or sows trained in protected contact (i.e., with a barrier between the trainer (outside the pen) and the animal (inside the pen)) can be positioned to enable the trainer to access an ear, a catheter, or the back of the animal. The trainer may also benefit from the fact that the animal will learn to expect where the (consistent) trainer presents the reinforcer and position her/himself (the animal) so that she/he can get the reinforcer as quickly as possible. If feasible, the target can be faded (made less and less obvious, e.g., by reducing the size), or gradually removed and replaced, e.g., by a hand signal.

LURING AND BAITING

Luring and baiting are methods where a prompt – a discriminative and often pleasant stimulus – is used to elicit a specific, desirable response. This prompt is presented prior to the behavioral response with an intentional location or mode of presentation that attracts the attention of the animal and prompts the behavior. In luring, the prompt is moved to elicit the behavior, whereas in baiting the pleasant stimulus is stationary, often placed in a location the trainer wants the animal to go.

Luring and baiting are often highly rewarding to the inexperienced trainer, because the animal shows the desired behavior immediately. Most animals can be lured by food or other desired stimuli to go to places or to do things. The lure is shown to the animal, moved to prompt the behavior, and the animal responds in whatever way is needed to get the lure. For example, if a food lure is presented in front of a dog's nose and then raised over the dog's head, most dogs will sit back almost automatically as the nose follows the lure. If the lure is taken downward, many dogs will lie down to get the lure. In the same way, a falconer can bait his glove with a dead mouse and the falcon will return to the glove to get the mouse. Properly used, a lure is removed when the animal shows the desired response and a reinforcer is applied. It is possible, though, to allow the animal to obtain the lure, which is always the case when baiting is used. Still, there is a fluid border between baits and lures. Luring and baiting have some drawbacks which the trainer should be aware of if they plan to use these techniques. The following points apply to both lures and baits. First, the animal will focus on the lure and not on her/his own behavior or the surroundings. Hence, there is a risk that the animal will not learn which specific behavior that leads to reinforcement, and she/he may not voluntarily present any behavioral responses without the lure being presented. Second, if the animal is lured into a potentially frightening environment, the animal may be startled by suddenly realizing that she/he is now on a ladder (climbing a ladder is a trained behavior, e.g., with police and rescue dogs) or in a very open area (which would potentially scare, e.g., rats and mice). Under such conditions the lure may even come to predict an unpleasant situation or even worse: the animal will associate the lure with both the unpleasant situation and the trainer. Third, the lure itself will soon become the cue or part of the cue; the discriminatory stimulus (Sd). As a result, the animal cannot perform the behavior without the trainer presenting the lure as a part of the cue. For example, the dog does not sit, when the owner says "sit," unless the owner has a piece of good food in his hand, because "food-in-hand" is a part of the "sit" cue. Interestingly, some animals will learn that if they ignore the first lure and withhold responding, the trainer will present a larger and more valuable lure. Thus, it is important that the trainer does not respond to increasing latency by offering a larger lure. Last, if the lure is suddenly withdrawn, or "wafted," the animal may be prone to chase the lure as it is removed, or, in the case of some predators, attack the trainer.

Lures, overall, can be a very effective training tool, but they should be used with caution and removed as soon as possible with a process called fading. During fading the lure is gradually removed, e.g., by making the lure less obvious or less valuable to the animal. Many trainers often begin training by luring an animal as it is an easy way to get the animal to show the behavior. However, luring should be used only to prompt behaviors if no other options are available. The process of luring, including the fading, should take place in three to ten trials (depending on the animal species and the individual). If luring is still used after eight to ten trials, the lure has not been used correctly and the trainer should stop training and review the training plan.

SHAPING

When a behavior is shaped, it simply means that the behavior is gradually built by reinforcing small steps towards the final goal, i.e., successive approximations towards the final behavior. When employing shaping to get a complex behavior it is of utmost importance to proceed with small enough steps to allow the animal to understand what the trainer wants. Be sure to split the goal behavior into small steps instead of asking for a more complex response – or as stated by Bailey and Breland, "be a splitter, not a lumper." Shaping may not always be the quickest and most efficient way to initiate behavior. However, when done correctly, it is probably the best demonstration of a trainer's knowledge as well as observational and mechanical skills.

HOW TO GET STARTED

Starting up animal training in the laboratory facility should start by choosing a reinforcer (often some kind of food) and making sure that the animal will eat the food when offered by the trainer.

Operant training should first be initiated when the animal accepts the primary reinforcer without latency. Moreover, it is of vital importance that the trainer is able to deliver the reinforcer fast and consistently without, e.g., spilling or dropping it. For some species, using tweezers or tongs may be advisable.

Foundation behaviors are behaviors that form the basis for most of the behaviors that we would like to train our laboratory animals to do. The four foundation behaviors we suggest to work with are "Stay calm," "Targeting," "Station," and "Gating." In laboratory animals, the most relevant foundation behaviors are the "Stay calm" and the "Targeting," which should be prioritized, if there is limited time for training.

The "Stay calm" behavior may be the most important behavior an animal can be trained to do and as such it should be the first behavior trained. The animal is reinforced for staying calm and relaxed, but fully aware of the trainer. Eye contact can be present, but it depends on the species and the position of the animal. For example, it is difficult for a mini pig to have eye contact with the trainer, if the mini pig is just in front of the trainer and the trainer is standing upright, then the "Stay calm" behavior is simply standing calmly in front of the trainer, waiting for the trainer to cue the next behavior. It is of course possible to actively train "Eye-contact" as a part of the "Stay calm" behavior; the relevance and feasibility must be decided by the trainer. Teaching the animal to stay calm results in controlled training sessions in which the trainer can better spot the target behavior and detect if the animal shows early signs of frustration. A calm animal can speed training a great deal and is worth the time and labor investment. A training session with an animal that does not show – as a baseline – calm behavior can be difficult to carry through with an optimal result.

Targeting is the behavior where the animal orients a specific body part (e.g., the snout) towards a stationary or moving object. It is imperative not to create a fear response when presenting the target for the first time or to introduce the target for the first time if the animal is not paying attention. The object can be moved to initiate a "follow-target" behavior that makes it possible to move the animal between pens/cages or into a procedure room fast and with great precision. It is also possible to keep the target immobile and have the animal standing still with, e.g., the snout calmly on the target for a shorter or longer duration. In this case, the target functions as a station.

For several reasons, it is advisable always to begin the training of an animal with these two foundation behaviors: the "stay calm" and the "targeting." First, if the animals are able to do these behaviors on cue, many challenging situations can be dealt with simply by asking the animal to stay calm or to follow the target. These behaviors will come to have a strong reinforcement history simply because they were started first and therefore have been in the animals' repertoire the longest. Behaviors with a long and strong reinforcement history often become a behavior that the animals will offer the trainer if the animal gets confused or misses a cue.

The third foundation behavior is the "Station" behavior. A station is part of the environment, a specific place in which the animal can be placed in a waiting position. It is important to realize that when an animal is calmly waiting on her/his station, the animal is still doing the correct behavior and should be regularly reinforced. When on station the animal will thus wait for the trainer to either reinforce the "Station" behavior or cue another behavior. It is vital not to postpone the reinforcement until just before the animal decides to leave; this may risk reinforcing "leaving station" behavior.

The fourth foundation behavior is "Gating" or "go from A to B." In some circumstances, being able to ask the animal to move from one point to another can be useful. It can be, for example, to move from one trainer to another or to move from one target (or station) to another. In most laboratory animals, this fourth foundation behavior may not be as relevant as the "Stay calm" behavior and the "Targeting," which should be prioritized.

INCORRECT RESPONSES AND UNWANTED BEHAVIORS

Any behavior other than the correct response is, in principle, an unwanted behavior. Moreover, a wide range of behaviors not relating to the training situation may also be categorized as unwanted behaviors. In this chapter, focus is solely on unwanted behaviors in human-animal interactions appearing in the training situation. Unwanted behaviors in animal-animal interactions (e.g. aggression in group-housed animals) are outside the scope of this chapter.

In the training setup, it may be useful to make a distinction between incorrect responses and unwanted behaviors. Incorrect responses denote the situations where the animal offers a behavior different from what the trainer asked for. Such a situation will occur from time to time during training sessions and the trainer must recognize its underlying cause and act accordingly. Unwanted behaviors, on the other hand, could be displays of aggression, frustration, or fear, consistently showing a specific incorrect response or repeatedly present behaviors not cued by the trainer.

INCORRECT RESPONSES

When an animal in the training situation shows an incorrect response, i.e., not the behavior that the trainer asked for, there may be several explanations. Perhaps the animal did not understand what the trainer asked for. In this case, the trainer may not have defined the criteria, have increased the criteria too fast (asking for too much) or the trainer has been inconsistent in cuing the behavior. Or perhaps the trainer has failed to correctly assess the antecedents and plan the training accordingly. Maybe the animal is not feeling well, maybe there are too many disturbances around the enclosure, or maybe she/he has the sun in her/his eyes – and the trainer failed to notice it. Or maybe the reinforcers used are simply not worth working for. One way to recognize that the training is heading in a wrong direction, is to be constantly aware of the animal's latency to respond. Often, when an animal shows one or more incorrect responses for the above-mentioned reasons, it has been preceded by an increasing latency to respond to the trainer's cues.

UNWANTED BEHAVIOR: AGGRESSION, FRUSTRATION, AND FEAR

Frustration and even aggression are often elicited by the behavior of inexperienced trainers failing to keep up a suitable rate of reinforcement. Fear may arise due to fear-eliciting stimuli in the environment at the beginning of the training. Therefore, the trainer's ability to identify antecedents that may elicit fear is important. Sudden, unexpected fear-eliciting stimuli may also occur during the training session. In such cases, knowing the animal, including body language, and facial expressions, is the key to identifying the changes in the animal's emotional and motivational state; changes that will predict possible unwanted behaviors. The trainer's task is to predict this and prevent it from happening.

UNWANTED BEHAVIOR: PERSISTENT OR PROGRESSING INCORRECT RESPONDING

Persistent or progressing incorrect responding may arise if the animal inadvertently and repeatedly has been reinforced for an incorrect response which the animal is now showing consistently. Such scenarios are most likely caused by inexperienced trainers and/or insufficient planning of the training.

Sometimes an animal will start showing progressing incorrect responding which counteracts the correct behavior, a phenomenon known as "instinctive drift." Breland and Breland (1961) described the phenomenon in which a learned response gradually drifts towards a more instinctive behavior associated with the used reinforcer to such an extent that the learned response does not happen at all. In their paper "The Misbehavior of Organisms," Breland and Breland (1961) present several examples of breakdowns of conditioned operant behaviors, including the example of the raccoon trained to pick up coins and deposit them in a five-inch metal box.[5] It was not difficult for the raccoon to learn to pick up the coins, but the raccoon had difficulties letting go of the coin; instead, the racoon was rubbing the coin, showing the typical "washing" behavior. With increasing demands such as performing many trials, having to repeat the behavior several times before getting reinforced (i.e., working on high ratios) or having to carry the coins for longer distances; the worse it became (personal communication 2019, R.E. Bailey). The raccoon's performance was acceptable as long as demands were not too high (personal communication 2019, R.E. Bailey). The Brelands concluded that the raccoon was actually showing "washing behavior," an instinctive behavioral pattern connected to foraging in raccoons. When working for food rewards, the raccoon tended to "drift" towards this instinctive behavioral pattern, which was different from and incompatible with the learned response.

No matter the cause for the incorrect response, the trainer needs to respond immediately and in a way that minimizes the risk of the animal getting frustrated, aggressive, or losing trust in the trainer. If the animal fails to show the correct response more than a couple of times, the trainer should stop the session and revise the antecedents, the training plans, and the training protocol.

How to Deal with Incorrect Responses

It is important to emphasize that incorrect responses most likely are because of trainer mistakes, and therefore incorrect responses should always result in the trainer assessing the session, evaluating possible trainer errors that may have led to the incorrect response.

Incorrect responses may be dealt with in various ways. If it is an isolated event the trainer may simply choose not to reinforce, use a few seconds to evaluate why the animal made a mistake and then proceed with the training in the most optimal way. Other options are the use of punishment procedures, reinforcement procedures focusing on differential reinforcement of alternative or incompatible behaviors (Table 6.1),[6–11] or by cueing another learned behavior such as the "Recall" or the "Stay calm." In punishment procedures the incorrect response will lead to either the removal of a positive stimulus (negative punishment) such as the trainer leaving and taking all the food with her/him – or the application of an aversive stimulus (positive punishment) such as physically correcting the animal or saying "No." Punishment procedures only communicate to the animal that the behavior is incorrect; it will not indicate which behaviors actually do result in reinforcement. Moreover, the trainer has no way of controlling the outcome, which may be aggression, suppression of behavior in general, or attempts to escape the trainer.[7] For these reasons, it may often be advisable to avoid the use of especially positive punishment as the main operant procedure in animal training. Another, and more advisable, strategy is to cue a well-trained behavior with which the animal is confident and then reinforce this behavior. These well-trained behaviors can be anything, depending on the animal and its reinforcement history.

If the incorrect responses have been mistakenly reinforced a number of times, they can be dealt with by employing a differential reinforcement schedule focusing on reinforcing other behaviors than the incorrect response or by the use of extinction (Table 6.1). Extinction is stopping a behavior by withholding reinforcement. The use of extinction implies that the incorrect behavior has previously been reinforced, most likely by mistake or lack of consideration and thorough planning. Depending on the level of previous reinforcement of the behavior, extinction may create an extinction burst, during which the animal will increase behavioral responding (see also Chapter 5). Such an extinction burst may be undesirable for the trainer.

HOW TO END A SESSION

Ending a session implies the risk that the animal perceives the trainer leaving as an unpleasant event in which case it may either function as a negative punishment and/or induce frustration or aggression. It is therefore imperative always to have a plan for when and how to end the session. In laboratory pigs, a technique that works well is to tell the animal to "go search" and simply throw a handful of reinforcers (pieces of apple, raisins, or the like) on the floor. When the pig is rooting, the trainer un-dramatically leaves. It is not necessary to end on a positive response. As incorrect responses are most likely due to trainer errors, the skillful trainer should not have any problems

TABLE 6.1

Techniques Used to Deal with Incorrect Responses (For Definitions of the Scientific Terms, Please Refer to Chapter 5)

Extinction[8,10]

Extinction

A weakening of a previously learned operant response as a consequence of lack of reinforcement of that response.

Punishment[7]

Non-Reward M

The non-reward marker (NRM) is a signal (e.g., the verbal "no") communicating that the behavioral response was not correct and consequently no reinforcement is given. In that sense the non-reward marker, however mild, is a positive punishment procedure.

Time Out

Time out is a negative punishment procedure. The trainer removes all possibilities to obtain a reinforcer as a consequence of an incorrect behavioral response. This may include the trainer turning their back to the animal or leaving the enclosure. The time-out thus may eliminate the trainers' possibilities to watch the animals' behavior and resume training, when feasible.

Reinforcement-Based Techniques

Differential Reinforcement of Other Behavior [7,10] (or Differential Reinforcement of Zero Responding[6,11]) (DRO)

Using a DRO implies reinforcing periods of time passing without the unwanted behavior being shown. For example, every time three minutes have passed without the dog barking, the dog is reinforced for zero responding (i.e., no barking). Hence, the DRO is a reinforcement procedure in which any response other than the unwanted behavior may be reinforced. The unwanted behavior will thus decrease, as it is never reinforced.

Differential Reinforcement of Alternative Behavior (DRA)[6]

If a dog is jumping on guests or a pig is continuously manipulating the hinges of the gate when it gets aroused, a DRA procedure would be to reinforce an alternative behavior such as sitting, lying down, eating, keeping four paws on the ground (in the dog's case), and rooting, making contact, exploring toys etc. (in the pig's case). In the case of the DRA the reinforced behavior does not have to be incompatible with the unwanted behavior.

Differential Reinforcement of Incompatible Behavior (DRI)[6]

If the trainer decides to focus on reinforcing one specific behavior, which is incompatible with the unwanted behavior (e.g., sitting in the case of the dog; the dog cannot jump, if s/he sits), it is traditionally referred to as a DRI (differential reinforcement of incompatible behavior).

Recall

It is possible to use the recall signal as a tertiary reinforcer presented as a consequence of the animal showing an incorrect response.

The recall can also be used simply as a cue to return to station or trainer even though the previous response was correct. Hence, the recall is not used only as a consequence of an incorrect response and the correct response to the recall signal should always be reinforced. Especially if animals are group housed, it is important to have a strong reinforcement history on the recall, as this behavior can be very important to have on cue.

getting all responses including the last one in the session correct. Insisting on ending on a positive response may drive the trainer to continue a training session that should otherwise have been ended, a situation which should be avoided.

CREATING A BALANCE

Sometimes animal trainers claim to be "positive animal trainers" or "positive reinforcement trainers" solely because they use operant conditioning with focus on positive reinforcement. However, grabbing a clicker and some treats are not enough to qualify as a "positive reinforcement trainer." The environment surrounding us and the animals will always be a mixture of pleasant stimuli and aversive stimuli (ranging from mild to extreme). In other words, there will always be both reinforcers and aversives in an animal's environment and animals spend a lot of time adjusting their behavior to optimize their welfare. During training, the antecedents likewise comprise a mixture of unpleasant, neutral, and pleasant stimuli and the animal will seek to tip the balance more towards the pleasurable stimuli, the reinforcers. "Positive animal training" is thus not simply working with positive reinforcers, focusing on positive reinforcement only; it is aiming to shift the balance towards as many pleasant stimuli in the form of pleasant antecedents as possible while reducing the aversives in the animal's environment to the highest possible extent. The trainer should therefore aim to reduce as much as possible the presence of unpleasant antecedents as positive reinforcement training can be antagonized by unpleasant events prior to or in the training situation (part of the antecedents) such as food restriction, stressful gating, lack of shelter to retreat to, separation of the animal from her/his group, or frustration because of a less skillful trainer.

ANIMAL TRAINING IN LABORATORY ANIMAL SCIENCE: CHALLENGES AND PERSPECTIVES

Working with laboratory animals, animal training should be an integral part of the experimental protocol. Efficiency and animal welfare might be enhanced if a trainer's knowledge and experience are included in protocol development. In general, the animal trainer should focus on the available options – and be creative and optimize within the limitations of the experimental protocol. If the animals have implanted catheters or if cage cleaning needs to be done daily, the training should be adapted to these restrictions. On the other hand, if increased animal welfare is the aim, the animal trainer should define what could be done to optimize the training and discuss these options with the principal investigator (PI) and the staff participating in the procedures with the purpose of finding new ways to bring training forward.

Several challenges and constraints exist with animal training in laboratory animals.

First, it is not always clear what the legal demands are. For example, in the EU it is stated that: "Establishments shall set up habituation and training programs suitable for the animals, the procedures and length of the project" (Directive 2010/63/EU, Annex III, 3.7). However, since the word "training" is not defined and the facility itself decides what is "suitable," the training applied is often limited, not necessarily adding to the welfare of the animal. One way to increase the use of animal training could be to include an animal trainer in ethical review boards and authorities granting licenses for animal experimentation. Evaluating the experimental protocol, the trainer would be able to suggest which procedures to train to and how.

Second, many researchers are reluctant to implement animal training as they are afraid that the training may affect the study and the data. It seems obvious, though, that the best data will be produced if the animals are as calm and confident as they can be. Moreover, certain procedures (such as taking multiple x-rays of a standing, calm, and un-sedated pig) can only be done by use of animal training as defined in this chapter.

A third obstacle is the lack of continuing education for laboratory animal caretakers and technicians. Training is a mechanical skill and good trainers need to understand the science behind the training and they need practice and experience. Hence, if animal training is to be successful, basic and continuing education of the staff is necessary. If the staff is not properly educated and able to perform the training efficiently and with good results, when animal training is initiated in the facility, the faulty conclusion may risk being that "animal training doesn't work" and it will be closed down again.

Fourth, several training constraints exist in the laboratory, the most common one being limited time. The animal may only be housed in the facility for a very short period. In some cases, the trainer may only have the institution's required acclimation period (one to two weeks) to train behaviors such as accepting an injection or standing still for a clinical examination. In other studies, the animals are housed for several weeks, but still there is no time set aside for animal training. Training the animals then may become something which can be done if there is time for it, when other tasks deemed more important have been done. In such cases, the results of the training will be sub-optimal and again, the conclusion may risk being that "animal training doesn't work." Allocating time for training may be a managerial challenge, especially if not all animal caretakers are able to train animals and a few people are asked to find time to do the training. It may be challenging for the remaining caretakers to willingly accept that some of their colleagues are spending time on training, leaving more of the less desirable tasks such as cleaning for the non-training staff.

Food restriction is another issue demanding the trainer to come up with either edible reinforcers accepted by the PI such as flavored water, gelatin, or ice cubes, or simply just pieces of a food item not containing any components interfering with the research parameters. Arranging training in these situations calls for good communication and trust between the PI and animal staff as well as an understanding of the complexity of the other parts work. Other issues may be, e.g., inflexible pen or cage systems and daily routines that cannot be combined with training.

Introducing animal training in the laboratory animal facility is possible in all animal species. All animals can be trained to perform any behavior they are capable of doing, given enough time and a skillful trainer. Larger animals such as dogs, pigs, and sheep are easily trained, whereas training the rodents often seems to be more challenging. Rodents are often kept in very high numbers and very little, if any, time is used to establish a meaningful human-animal relation on which training can be based. Until this attitude towards rodents changes, the benefits of training may not be available to mice and rats. Luckily, it seems that an increased awareness is rapidly growing in recent years. As more and more studies are published in which animals have been trained to voluntarily cooperate, it will be increasingly clear to legal authorities that animal training is a Refinement with tremendous impact on animal welfare. Coincidently, researchers will hopefully realize that unstressed and confident animals are better models and funding will be raised to ensure the better model. Training animals to cooperate during experimental procedures should be the ultimate Refinement goal and the technology is there – we just need to implement it.

ACKNOWLEDGMENT

The "Bailey-Breland proverbs" mentioned in this chapter have all been coined by Marian Breland-Bailey and Robert E. (Bob) Bailey.

REFERENCES

1. Skinner, B. F. *The Behavior of Organisms: An Experimental Analysis* (New York: Appleton-Century-Crofts, Inc., 1938).
2. Thorndike, E. L. *Animal Intelligence: Experimental Studies* (New York: The Macmillan Company, 1911).
3. Pavlov, I. P. In Anrep, G. V. (trans., ed.) *Conditioned Reflexes: An Investigation of the Physiological Activity of the Cerebral Cortex* (Oxford: Humphrey Milfold, Oxford University Press, 1927).

4. Feng, L. C., Howell, T. J. & Bennett, P. C. How clicker training works: Comparing reinforcing, marking, and bridging hypotheses. *Applied Animal Behaviour Science* **181**, 34–40 (2016). doi:10.1016/j.applanim.2016.05.012

5. Breland, K. & Breland, M. The misbehavior of organisms. *American Psychologist* **16**, 681–684 (1961). doi: 10.1037/h0040090

6. Martin, G. & Pear, J., Differential reinforcemetn procedures ot decrease behavior. In *Behavior Modification: What It Is and How to Do It* Ch. 14, (New York: Routledge - Taylor and Francis Group, 2019).

7. Chance, P., Punishment, In *Learning and Behaviour* Ch. 7 (Belmont, CA: Wadsworth, Cengage Learning, 2009).

8. Chance, P., Reinforcement. In *Learning and Behaviour* Ch. 5 (Belmont, CA: Wadsworth, Cengage Learning, 2009).

9. Chance, P., Schedules of reinforcement. In *Learning and Behaviour* Ch. 6 (Belmont, CA: Wadsworth, Cengage Learning, 2009).

10. Grant, L. & Evans, A., Decreasing responding; Extinction, DR0 and DRI. In *Principles of Behavior Analysis* Ch. 3 (New York: HarperCollins College Publishers, 1994).

11. Vollmer, T. R. & Iwata, B. A. Differential reinforcement as treatment for behavior disorders: Procedural and functional variations. *Research in Developmental Disabilities* **13**, 393–417 (1992). doi: 10.1016/0891-4222(92)90013-v

7 The Zebrafish

Isabel Fife-Cook, Christine Powell, and Becca Franks

CONTENTS

INTRODUCTION

Zebrafish (*Danio rerio*) are a freshwater teleost fish native to the Ganges and Brahmaputra river basins and the surrounding floodplains across northeastern India, Bangladesh, and Nepal.[1] They are also currently one of the most popular model organisms in biomedical research. It is estimated that over 3000 institutions conduct research involving *D. rerio*, with usage expected to rise in the coming years following the advent of CRISPR-*cas9* gene editing.[2]

Numerous efforts have been made to establish standardized husbandry practices for laboratory zebrafish.[3,4] Nevertheless, ongoing refinement of these guidelines is necessary as our knowledge of zebrafish behavior and preferences continues to progress. Thus far, the majority of zebrafish welfare research has focused on areas such as euthanasia, anesthesia, and other attempts to mitigate the aversive aspects of life in the laboratory.[5,6] Recently, however, information is accumulating regarding their preferences for structural complexity,[7] potential desire for free-choice exploration opportunities,[8] and their range of social dynamics.[9] While more work is needed, the existing literature already contains many indications of promising ways to elevate and evaluate zebrafish welfare.

As with all considerations of welfare, it is useful to begin with careful consideration of biological health, what life would be like in the wild, and affective states (Figure 7.1).[10] Current welfare guidelines for laboratory zebrafish mainly aim to cultivate physiologically sound specimens and rarely, if ever, address the individual's psychological health or the naturalness of the behavior. Although maintaining healthy animals is an indisputably necessary welfare goal, good biological health is not in and of itself sufficient to ensure a good life, especially when the animal's psychological, environmental, and relational needs are not addressed.[11] Developing an animal-first approach to zebrafish welfare therefore requires prioritizing these additional dimensions and embracing welfare as a way to enhance the scientific value of the research.

THE BASICS: BIOLOGICAL HEALTH

Maintaining high standards of management and husbandry for zebrafish includes monitoring water quality, providing appropriate nutrition, practicing disease and genetic management, completing

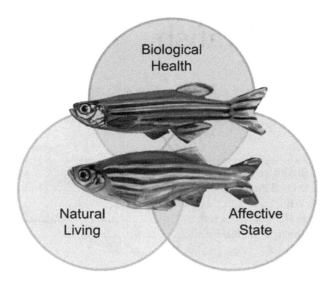

FIGURE 7.1 Top: healthy male (narrow body). Bottom: healthy female (rounded body). For complete zebrafish welfare, it is necessary to consider all three spheres of concern: biological health, natural living and behavior, and affective states and psychological experience more generally.

regular health assessments, and maintaining regulatory compliance (for detailed instructions regarding water quality maintenance see Reed and Jennings, 2011[4] and Avdesh et al., 2012[12]). It is also necessary that all staff receive adequate training to assess zebrafish health, including knowledge of both behavioral and physiological symptoms of disease and distress (see Table 7.1).[3]

Zebrafish analgesia (pain relief), anesthesia (loss of sensation and awareness), and euthanasia (facilitated mortality) are all important additional considerations, and research in these areas is ongoing. Regardless of the procedure administered, however, thorough training of caretakers on level of sedation (e.g., ventilation rates, posture, and response to stimuli) is necessary. A complete list of recommended procedures can be found as outlined by the American Veterinary Medical Association (2020).[13,14]

ENVIRONMENTAL CONDITIONS AND NATURAL BEHAVIOR

Housing animals in environments that they find to be aversive[16] or under-stimulating[17,18] denies them good welfare as it can cause abnormal behavior (including stereotypic behavior), increased sensitivity to stress, impaired brain function, lethargy, increased aggression, decreased growth rates, and decreased immunity.[11] In addition to the moral imperative of housing sentient animals adequately, these factors can adversely influence the results of experimental studies and reduce the scientific validity of the data. In contrast, including biologically relevant elements of a species' natural environment can have a positive impact on animal welfare and improve the quality of the research.[10,20,21]

Compared to the information about their behavior in laboratory settings, information regarding the natural behavior of zebrafish in the wild is limited. As such, the accepted standards for "normal" behavior could be biased towards how zebrafish behave in restrictive laboratory settings, inhibiting the ability of caregivers to identify and resolve welfare problems. Paying close attention to what is known about behavior in natural or semi-natural settings and distinguishing it from patterns of behavior in more artificial environments is therefore necessary.

ZEBRAFISH LIFE IN THE WILD

Wild *Danio rerio* are typically found in slow-moving or stagnant pools, manmade lakes, ponds, and irrigation channels constructed for agricultural use,[22] typically with overhanging vegetation[23]

TABLE 7.1
Indicators of Biological Health

Normal Behavior (Smith, 2014)[15]	Abnormal Behavior
Occupies entire water column.	Hanging around the bottom, difficulty maintaining balance, and position in the water column.
Moderate activity levels.	Erratic bursts of activity.
	Lethargy.
	Increased respiratory rate.
Infrequent displays of aggression.	Frequent or constant aggression.
Normal Appearance	**Abnormal Appearance**
Normal external morphology.	Abnormal external morphology:
	• Clamped or frayed fins.
	• Hypo- or hyper-pigmentation.
	• Protruding or missing scales.
	• Air bubbles in eye, fins, or gills.
	• Increased mucus production on skin or gills.
	• Corneal opacity or cataracts.
	• Coelomic distension.
	Rapid weight loss.
Normal Performance	**Abnormal Performance**
Uniform egg size/acceptable hatchability.	Variable egg and embryo size/lower hatchability than expected.
Acceptable survival rates.	Lower survival than expected.
Acceptable growth rate/reaches sexual maturity in appropriate amount of time.	Variable size within spawning/does not reach sexual maturity in acceptable amount of time.

(Figure 7.2). Zebrafish can occupy the entire water column and spend time in open water as well as amongst aquatic vegetation[22] in addition to a wide range of temperatures, pH levels, and turbidity levels.[23] Zebrafish also have flexible group sizes, with field observations noting shoals (cohesive groupings) of six to 300 individuals.[24]

RECREATING THE NATURAL HABITAT IN THE LABORATORY

As a highly social species, the most basic element of naturalistic housing for zebrafish is appropriately sized social groups and adequate space to accommodate the size of the social group.[25] Avoiding both social isolation and overstocked conditions are critical.[19]

Beyond social considerations, the presence and type of substrate also play an important role in the lives of zebrafish. Substrate allows zebrafish to engage in natural foraging and mating behavior and has been found to lower activity levels when provisioned in laboratory environments.[26] Moreover, zebrafish consistently prefer areas with substrate over barren environments, choose gravel substrates over silt, and even prefer simple images of gravel over barren tank bottoms.[7,27] As such, the use of gravel substrate is likely to improve zebrafish welfare and, when possible, real gravel (as opposed to stickers or inserts designed to resemble gravel) is strongly encouraged in order to provide them with the opportunity to engage in natural foraging and mating behavior.

In the wild, zebrafish also spend a substantial amount of time amongst vegetation[22] and the introduction of plants in captivity has been shown to have positive effects on welfare.[28] Aquatic plants thought to be similar to those found in the wild include vallis (*Vallisneria spp.* including *V. spiralis*, *V. elongata* and *V. tortifolia*) and water trumpet (*Cryptocoryne wendtii*).[29] Across several studies, zebrafish have shown a strong preference for areas with artificial plants[19,28,30,31] and the presence of artificial plants has been found to decrease anxiety behaviors[28] (see Figure 7.2).

FIGURE 7.2 a) Zebrafish life in the wild consists of complex waterways, containing currents, ample vegetation above and under water, a variety of substrates, and a diversity of animal life; b) naturalistic housing in the laboratory could contain real substrate (gravel or sand), live plants, logs or rocks, dark walls, depth variety (i.e., sloped gravel), full-spectrum overhead light, simulated gradual dawn and dusk light cycle, and invertebrates such as snails and freshwater shrimp; c) restricted housing in the lab could contain images of substrate, fake plants, dark walls, full spectrum overhead light, and simulated gradual dawn and dusk light cycle; d) standard laboratory housing consists of a barren tank with transparent blue walls on all sides (including the bottom), and ambient room lighting, potentially with sudden on/off light cycle.

The light cycle is another salient variable for laboratories to consider. Zebrafish show diurnal activity patterns that are influenced by the photoperiod,[1] and the onset of the light cycle may play a key role in instigating spawning behaviors.[27] Moreover, gradual transitions in lighting may be particularly beneficial to fish, as sudden light transitions have been found to induce a startle response.[32]

Finally, zebrafish have been found to prefer the deep end of a tank, potentially as a way to protect themselves from aerial attacks.[33] It is possible, however, that this depth preference is a result of captivity-induced stress rather than an innate preference. By equipping tanks with varying water-depth, caretakers can provide fish with more choice and control while also simulating their natural environment. Depth differentials can be achieved by using layers of substrate, rocks, or other inert structures to create a slope against the tank walls to imitate the bank of a river (Figure 7.2).

Accounts of zebrafish behavior displayed under naturalistic conditions reveal novel patterns compared to what is known under standard laboratory conditions, which suggest that their behavioral repertoire may be greater than previously understood. For example, when small groups of zebrafish were housed in large (110 L) naturalistic tanks, fish were recorded digging a space underneath an artificial plant base, swimming in and out of the area along the gravel, circling around each other in close proximity, and displaying quivering behavior characteristic of spawning.[25] It is possible that such activities represent nesting behavior, though more research is necessary to determine the nature of this activity.

Similarly, when housed in large, 1100 liter tanks furnished with substrate and artificial plants, wild-caught zebrafish were found to conduct pair breeding rather than group breeding, previously thought to be the norm.[34] Neither digging nor pair-breeding behavior have been reported under traditional laboratory conditions, indicating that these behaviors may be dependent on tank size and substrate availability, underscoring the need for more research on zebrafish behavior in naturalistic conditions (Figure 7.3).

PSYCHOLOGICAL AND EMOTIONAL WELL-BEING

Emotional experience is a central component of animal welfare[35–37] and is now considered to be an adaptive mechanism defined as "affective responses to an event manifested through physical

FIGURE 7.3   Observations of several different tanks of naturalistically housed zebrafish revealed a previously unreported zebrafish behavior that may be a form of nesting. Each morning around daybreak (when zebrafish typically breed), zebrafish were observed swimming repeatedly in and out of a cavity that formed underneath the solid base of an artificial plant. The activity was marked by an absence of aggression, high rates of participation (all fish in the tank would congregate around the base), and high fish densities (many fish squeezing into the small space at the same time; see Graham, et al., 2018b for a link to a video of this behavior).

changes"[35] (see Table 7.2 for examples of assessing affective states in zebrafish). Emotions help animals avoid harmful situations while also directing them towards desirable resources and rewards.[38] Moreover, beyond the historic focus on negative emotions such as fear, anxiety, pain, and aggression, positive emotions such as interest, caring/affiliation, and amusement, are increasingly recognized as more than mere luxuries; they are now thought to be central to understanding welfare and well-being. In humans, for example, positive experiences have been found to protect against boredom, languishing, and emotional distress; and positive emotions strengthen long-term well-being

TABLE 7.2

Indicators of Psychological Health

Assessment Approach	Indicators of Positive Psychological Health	Indicators of Negative Psychological Health
Behavioral.	Engaging in natural behaviors such as foraging, shoaling, low levels of aggression.[61] Affiliative behavior.[8] Information-gathering (exploration).[8] Heightened shoaling.[9]	Response to alarm pheromone:[62] • Reduced exploration. • Erratic movements. • Freezing behavior. Depression:[63] • Reduced reward behavior (e.g., conditioned place preference). • Hypophagia (reduced food intake). • Hypolocomotion (reduced baseline activity). • Aberrant sleep phenotypes. • Reduced sexual activity. • Cognitive deficits.
Physiological.	Elevated levels of dopamine, serotonin and isotocin.[61]	• Elevated cortisol levels. • Activated HPI (hypothalamic pituitary adrenal axis).

and resiliency, and even improve biological outcomes.[39] Crucially, similar patterns are now being identified across the animal kingdom[17,40,41] including fish.[42]

As a highly gregarious species, it is possible that zebrafish may feel vulnerable to predation if they are alone. Consistent with this possibility, isolated zebrafish display various signs of distress and dysfunction.[43–45] As such, isolation should be avoided at all costs,[46] but if it is unavoidable, hiding places in the form of structures, dark-walled containers, or plants may mitigate stress.[28,47] Previous research has shown that mere visual contact with conspecifics is rewarding to zebrafish,[46,48] suggesting a possibility that placing transparently walled tanks next to each other could minimize some of the stress of social isolation. However, it is unknown whether visual contact with conspecifics supersedes their preference for dark tank walls – more work is needed to determine the relative value of these parameters and how they affect zebrafish welfare. Beyond the presence/absence of conspecifics, the composition of the shoal is also likely to matter. At a minimum, mixed-sex groups are recommended as sexually mature females who are not exposed to male pheromones may fail to release their eggs and risk becoming "eggbound", a condition that can be fatal.[1]

Cognitive stimulation plays an important role in the welfare of many captive animals.[49–51] Zebrafish are known to have high cognitive capacities,[52] including the ability to learn and retain information involved in a variety of tasks, such as associating a food reward with a specific color,[53] using a tunnel maze to access a group of conspecifics,[54] and engaging in social learning.[55] As such, zebrafish may also require cognitive enrichment opportunities in order to thrive. Recent research confirms this possibility, demonstrating that zebrafish are often neophilic, meaning that they will readily engage with novel stimuli,[56] and explore novel spaces when given the free-choice opportunity to do so.[8] More work is needed to determine the long-term effects of providing zebrafish with their preferences for cognitive stimulation vs. not.

In the meantime, caretakers can carefully experiment with providing zebrafish with free-choice opportunities to engage in beneficial cognitive enrichment activities. As with all forms of enrichment, however, it is critical that the interventions are presented to the animals in such a way that they are able to have control over their decision to engage vs. ignore the opportunity.[4,57] For example, in larger tanks, part of the space could be designated as a novel area, separated with a divider that could be raised or lowered to provide intermittent exploration and learning opportunities (see Graham et al., 2018a[8] for a rudimentary application of this concept). It should be noted, however, that simply increasing environmental complexity is not necessarily synonymous with providing biologically relevant environmental improvement.[20,58] For example, consistent with the idea that increased complexity alone is not sufficient to promote welfare, zebrafish did not prefer and showed no signs of benefit from being housed in "enriched" tanks with glass rods that increased structural and spatial complexity.[58] Consideration of the natural habitat and the motivational experience of zebrafish are also critical.[59]

In addition to cognitive stimulation, animals are also likely to benefit from opportunities to make choices and have control.[35,40,59,60] For example, zebrafish could be given the opportunity to decide between two or more substrate types, areas with full vs. partial overhead cover (particularly in small group sizes), and various materials for spawning sites and shelter.[7,19] As with all forms of enrichment, however, it is worth considering equal access because aggression can break out if preferred resources are subject to monopolization by a few individuals.

SPECIAL CONSIDERATIONS

STRAINS

Significant differences have been noted between strains of recently caught zebrafish (i.e., "wild-type") and those who have lived in captivity for generations (i.e., TMI, SH, and AB strains).[1] Strains with a longer history of living in captivity show behavioral and physical signs of domestication. Domesticated strains grow faster and to a larger weight than wild zebrafish, likely due to higher feed intake in captivity.[1,64] In addition, zebrafish strains with a longer history of living in captivity differ in their collective social behaviors, engage in less bottom-dwelling, and present a muted startle response when compared with wild-type counterparts.[65,66]

However, it should be noted that not all behavioral differences between strains can be directly attributed to the forces of domestication. Among wild-type strains, there is large behavioral variation.[67] Wright (2003) found that boldness in wild-type zebrafish was correlated with where the fish were originally captured, as measured by latency to approach a novel object. Just as they have varying behaviors, different strains may also have varied preferences and needs that should be accommodated – perhaps analogous to the needs of different dog breeds. The present evidence suggests that wild-type and recently captive strains are more sensitive to predation threats and future research should address strain differences in human-animal interactions, structural enrichment, and social housing.

HUMAN-ANIMAL INTERACTION

Human-animal interaction can have both positive and negative effects on the animal's psychological welfare (see Chapter 3). Anecdotal experience with laboratory zebrafish indicates a high level of sensitivity to human activity, suggesting that technicians may be able to influence the welfare of zebrafish by changing small aspects of their interactions.

For instance, procedures that involve removing zebrafish from a familiar environment (such as involuntary transport to a new environment) may induce high levels of stress and fear that have a negative impact on both animal welfare and research validity.[68] To avoid these negative experiences, technicians should take care to maximize submersion and minimize transport time (e.g., capturing fish in the transport container underwater to ensure sustained submersion) and avoid netting, a highly stressful transportation method involving oxygen deprivation.[69] Using a scoop rather than a net has been found to mitigate some of the stress of transport in other species.[70] Additionally, considering that zebrafish prefer a dark environment in situations of stress,[47] the use of a dark, opaque container may also mitigate the negative effects of transportation.

Everyday husbandry stress may also be minimized by instructing all personnel to behave calmly and predictably around the fish – predictability has been found to decrease the stress reaction to aversive events.[71] Adding predictability to husbandry procedures such as tank cleaning may also enable animals to understand and better cope with these otherwise uncontrollable events. Further, introducing zebrafish to common husbandry tools while they are not in use (siphon, scrub brush, net), may be a useful way to habituate them to these items and mitigate some stress of their procedures.

On a more positive note, laboratory technicians can also take steps to facilitate positive associations with human interaction. Positive reinforcement training can be used to teach captive animals to voluntarily participate in necessary husbandry events. Training increases the predictability and therefore cognitive control animals have, mitigating the stress of these procedures.[72] As well as curtailing some of the negative aspects of captive life, training has the potential to serve as a form of enrichment for zebrafish. Many mammals voluntarily interact with learning devices, and cognitive activities such as learning have been associated with signs of positive affect.[49] Because of zebrafishes' cognitive abilities and willingness to engage in learning activities, training or even simple human interaction may be rewarding in itself for this species. More research should be done to investigate these exciting possibilities (see Chapter 1 for more information on approaches to human-animal interactions to improve welfare).

RESTRICTIVE TYPES OF RESEARCH

Traditional zebrafish researchers may question the practicality of implementing welfare-friendly housing and how such considerations would affect the reliability and reproducibility of their results. For example, current standards require that toxicology studies house zebrafish in tanks with inert water-flow, minimal microbial growth, and complete visibility of the fish for behavioral observation.[58] These attempts at standardization and control, however, do not account for the internal tank differences that can dramatically alter results, e.g., stress and dominance rank influences,[73,74] nor do they consider the negative impacts of restricting the zebrafish's natural ability to find homeostasis through behavioral regulation. Moreover, barren and inert conditions are not a way of removing environmental influences, they are just a very particular (and peculiar) type of environment.

As such, the current laboratory standards introduce limiting conditionalities regarding the inferential potential of the research. In other words, standard laboratory research is sampled from an unrealistic population and while it may be relevant to that narrow population, it should not be assumed to be relevant to the diverse populations that constitute the real world.[75] Thus, while there are certainly legitimate concerns regarding the practical implementation of zebrafish-centric methods, stringent laboratory standardization procedures may be precisely what undermines scientific validity.[76] And, if the research is not valid, it is not worth any investment, even if that investment is minimized by cost-effective housing practices.

CONCLUSION

The rapidly growing popularity of *D. rerio* as model organisms in biomedical research necessitates a careful and thorough weighing of their welfare under laboratory conditions. An animal-first approach to welfare assessment and implementation requires laboratory managers, veterinarians, researchers, and technicians to prioritize the animals' well-being and work collaboratively towards providing laboratory animals with excellent healthcare, species-appropriate housing, and the opportunity to lead rich and stimulating lives. Standardization of welfare guidelines and husbandry procedures is a necessary step in achieving animal-first welfare goals.

The ability to display natural behavior and interact with elements of the natural environment (such as live and artificial plants, gravel, and hiding places) is crucial because it allows animals to engage in important natural behavior and can mitigate the detrimental effects of under-stimulation and monotony. Naturalistic elements are necessary components of laboratory housing and have the potential to improve research validity. Future research on the natural behavior of zebrafish should similarly be prioritized in order to establish accurate baseline behavior.

Group composition also has a significant impact on welfare. To minimize the negative effects of inappropriate grouping (such as increased aggression) fish should be grouped in the recommended sex ratios/numbers in sufficient space. Additionally, laboratory personnel can help improve zebrafish welfare by mitigating negative experiences associated with human-animal interaction: use of analgesics and anesthetic when appropriate, reducing stress associated with transport, and encouraging positive interactions.

We recognize that elements of the recommended environmental parameters and husbandry guidelines presented in this chapter differ substantially from current standard laboratory protocol. However, we also recognize that standard laboratory housing for zebrafish differs greatly from what is known about their lives in the wild. While financial and research-based restrictions are unavoidable, considering the animal's well-being remains the primary priority of animal-centric welfare and should play a central role in shaping the direction of resource allocation and experimental design going forward.

REFERENCES

1. Spence, R., Gerlach, G., Lawrence, C. & Smith, C. The behaviour and ecology of the zebrafish, *Danio rerio*. *Biological Reviews* **83**, 13–34 (2008).
2. Lidster, K., Readman, G. D., Prescott, M. J. & Owen, S. F. International survey on the use and welfare of zebrafish *Danio rerio* in research. *Journal of Fish Biology* **90**, 1891–1905 (2017).
3. Lawrence, C. New frontiers for zebrafish management. *Methods in Cell Biology* **135**, 483–508 (2016).

4. Reed, B. & Jennings, M. *Guidance on the Housing and Care of Zebrafish Danio rerio.* (Southwater: Royal Society for the Prevention of Cruelty to Animals, 2011).

5. Deakin, A. G., Spencer, J. W., Cossins, A. R., Young, I. S. & Sneddon, L. U. Welfare challenges influence the complexity of movement: Fractal analysis of behaviour in zebrafish. *Fishes* 4(1), 8–16 (2019).

6. Valentim, A. M. et al. Euthanizing zebrafish legally in Europe: Are the approved methods of euthanizing zebrafish appropriate to research reality and animal welfare? *EMBO Reports* 37, 264–278 (2016).

7. Schroeder, P., Jones, S., Young, I. S. & Sneddon, L. U. What do zebrafish want? Impact of social grouping, dominance and gender on preference for enrichment. *Laboratory Animal* 48, 328–337 (2014).

8. Graham, C., von Keyserlingk, M. A. G. & Franks, B. Free-choice exploration increases affiliative behaviour in zebrafish. *Applied Animal Behavioural Science* 203, 103–110 (2018).

9. Franks, B., Graham, C. & von Keyserlingk, M. Is heightened-shoaling a good candidate for positive emotional behavior in zebrafish? *Animals* 8, 152 (2018).

10. Fraser, D., Weary, D., Pajor, E. & Milligan, B. A scientific conception of animal welfare that reflects ethical concerns. *Animal Welfare* 6(19), 187–205 (1997).

11. Mellor, D. J. Moving beyond the "Five Freedoms" by updating the "Five Provisions" and introducing aligned "Animal Welfare Aims". *Animals* 6, 59 (2016).

12. Avdesh, A. et al. Regular care and maintenance of a zebrafish (*Danio rerio*) laboratory: An introduction. *JoVE (Journal of Visualized Experiments)* 69, e4196 (2012).

13. Leary, S. & Johnson, C. L. *AVMA Guidelines for the Euthanasia of Animals*: 2020 Edition. Members of the Panel on Euthanasia AVMA Staff Consultants.

14. Collymore, C. Anesthesia, analgesia, and euthanasia of the laboratory zebrafish. In *The Zebrafish in Biomedical Research* (London: Academic Press, 2020), pp. 403–413.

15. Smith, S. A. Welfare of laboratory fishes. *Laboratory of Animal Welfare* 301–311 (2014). doi: 10.1016/B978-0-12-385103-1.00017-8

16. Dawkins, M. S. From an animal's point of view: Motivation, fitness, and animal welfare. *Behavioural. Brain Sciences* 13, 1–61 (1990).

17. Burn, C. C. Bestial boredom: A biological perspective on animal boredom and suggestions for its scientific investigation. *Animal Behaviour* 130, 141–151 (2017).

18. Dawkins, M. S. *Behavioural Deprivation: A Central Problem in Animal Welfare* 20(3–4), 209–225 (1988).

19. Kistler, C., Hegglin, D., Würbel, H. & König, B. Preference for structured environment in zebrafish (*Danio rerio*) and checker barbs (*Puntius oligolepis*). *Applied Animal Behaviour Science* 135, 318–327 (2011).

20. Newberry, R. C. Environmental enrichment: Increasing the biological relevance of captive environments. *Applied Animal Behaviour Science* 44(2–4), 229–243 (1995).

21. Yeates, J. W. Naturalness and animal welfare. *Animals* 8, 53 (2018).

22. Spence, R. et al. The distribution and habitat preferences of the zebrafish in Bangladesh. *Journal of Fish Biology* 69, 1435–1448 (2006).

23. Engeszer, R. E., Patterson, L. B., Rao, A. A. & Parichy, D. M. Zebrafish in the wild: A review of natural history and new notes from the field. *Zebrafish* 4, 21–40 (2007).

24. Suriyampola, P. S., Sykes, D. J., Khemka, A. & Shelton, D. S. Water flow impacts group behavior in zebrafish (*Danio rerio*). *Behavioral. Ecology* 28, 94–100 (2017).

25. Graham, C., von Keyserlingk, M. A. G. & Franks, B. Zebrafish welfare: Natural history, social motivation and behaviour. *Applied Animal Behaviour Science* 200, 13–22 (2018).

26. von Krogh, K., Sørensen, C., Nilsson, G. E. & Øverli, Ø. Forebrain cell proliferation, behavior, and physiology of zebrafish, *Danio rerio*, kept in enriched or barren environments. *Physiology Behavior* 101, 32–39 (2010).

27. Spence, R., Ashton, R. & Smith, C. Oviposition decisions are mediated by spawning site quality in wild and domesticated zebrafish, *Danio rerio. Behaviour* 144, 953–966 (2007).

28. Collymore, C., Tolwani, R. J. & Rasmussen, S. The behavioral effects of single housing and environmental enrichment on adult zebrafish (*Danio rerio*). *Journal of American Association Laboratory Animal Science* 54, 280–5 (2015).

29. Lee, C. J., Paull, G. C. & Tyler, C. R. Effects of environmental enrichment on survivorship, growth, sex ratio and behaviour in laboratory maintained zebrafish *Danio rerio. Journal of Fish Biology* 94(1), 86–95 (2018).

30. Delaney, M. et al. Social interaction and distribution of female zebrafish (*Danio rerio*) in a large aquarium. *Biology Bulletin* 203, 240–241 (2002).

31. DePasquale, C., Fettrow, S., Sturgill, J. & Braithwaite, V. A. The impact of flow and physical enrichment on preferences in zebrafish. *Applied Animal Behaviour Science* **215**, 77–81 (2019).
32. Emran, F., Rihel, J. & Dowling, J. E. A Behavioral assay to measure responsiveness of zebrafish to changes in light intensities. *Journal of Visualized Experiments* **20**, e923 (2008).
33. Blaser, R. E. & Goldsteinholm, K. Depth preference in zebrafish, *Danio rerio*: Control by surface and substrate cues. *Animal Behaviour* **83**, 953–959 (2012).
34. Hutter, S., Penn, D. J., Magee, S. & Zala, S. M. Reproductive behaviour of wild zebrafish (*Danio rerio*) in large tanks. *Behaviour* **147**, 641–660 (2010).
35. Boissy, A. et al. Assessment of positive emotions in animals to improve their welfare. *Physiology & Behavior* **92**, 375–397 (2007).
36. Špinka, M. Social dimension of emotions and its implication for animal welfare. *Applied Animal Behaviour Science* **138**, 170–181 (2012).
37. Webb, L. E., Veenhoven, R., Harfeld, J. L. & Jensen, M. B. What is animal happiness? *Annals of the New York Academy Sciences* **1438**, 62–76 (2019).
38. Fraser, D. & Duncan, I. J. H. 'Pleasures','Pains' and animal welfare: Toward a natural history of affect. *Animal Welfare* **7**, 383–396 (1998).
39. Fredrickson, B. L. The role of positive emotions in positive psychology: The broaden-and-build theory of positive emotions. *American Psychologist* **56**, 218–226 (2001).
40. Franks, B. & Higgins, E. T. Effectiveness in humans and other animals: A common basis for well-being and welfare. In Olson, J. M. & Zanna, M. P. (Eds.), *Advances in Experimental Social Psychology.* Vol. 46, 285–346 (Elsevier Academic Press) (2012).
41. Špinka, M. & Wemelsfelder, F. Environmental challenge and animal agency. In Appleby, M. C. (Ed.), *Animal Welfare* (CAB International, 2018), pp. 39–55.
42. Fife-Cook, I. & Franks, B. Positive welfare for fishes: Rationale and areas for future study. *Fishes* **4**, 1–14 (2019).
43. Shams, S., Amlani, S., Buske, C., Chatterjee, D. & Gerlai, R. Developmental social isolation affects adult behavior, social interaction, and dopamine metabolite levels in zebrafish. *Developmental Psychobiological* **60**, 43–56 (2017).
44. Shams, S., Chatterjee, D. & Gerlai, R. Chronic social isolation affects thigmotaxis and whole-brain serotonin levels in adult zebrafish. *Behavioural Brain Research* **292**, 283–287 (2015).
45. Ziv, L. et al. An affective disorder in zebrafish with mutation of the glucocorticoid receptor. *Molecular Psychiatry* **18**, 681–691 (2013).
46. Velkey, A. J. et al. High fidelity: Assessing zebrafish (*Danio rerio*) responses to social stimuli across several levels of realism. *Behavioural Processes* **164**, 100–108 (2019).
47. Serra, E. L., Medalha, C. C. & Mattioli, R. Natural preference of zebrafish (*Danio rerio*) for a dark environment **32**, 1551–1553 (1999).
48. Al-Imari, L. & Gerlai, R. Sight of conspecifics as reward in associative learning in zebrafish (*Danio rerio*). *Behavioural Brain Research* **189**, 216–9 (2008).
49. Franks, B. Cognition as a cause, consequence, and component of welfare. In Mench, J. A. (Ed.), *Advances in Agricultural Animal Welfare: Science and Practice*, 3–24 (Woodhead Publishing, 2018).
50. Higgins, E. T. *Beyond Pleasure and Pain: How Motivation Works* (Oxford University Press) (2012).
51. Meehan, C. L. & Mench, J. A. The challenge of challenge: Can problem solving opportunities enhance animal welfare? *Applied Animal Behaviour Science* **102**, 246–261 (2007).
52. Williams, F. E., White, D. & Messer, W. S. A simple spatial alternation task for assessing memory function in zebrafish. *Behavioural Processes* **58**, 125–132 (2002).
53. Mueller, K. P. & Neuhauss, S. C. F. Automated visual choice discrimination learning in zebrafish (*Danio rerio*). *Journal of Integrative Neuroscience* **11**, 73–85 (2012).
54. Gómez-Laplaza, L. M. & Gerlai, R. Latent learning in zebrafish (*Danio rerio*). *Behavioural Brain Research* **208**, 509–515 (2010).
55. Zala, S. M. & Määttänen, I. Social learning of an associative foraging task in zebrafish. *Naturwissenschaften* **100**, 469–72 (2013).
56. Lucon-Xiccato, T. & Dadda, M. Assessing memory in zebrafish using the one-trial test. *Behavioural Processes* **106**, 1–4 (2014).
57. Näslund, J. & Johnsson, J. I. Environmental enrichment for fish in captive environments: Effects of physical structures and substrates. *Fish Fish* **17**(1), 1–30 (2014).
58. Wilkes, L., Owen, S. F., Readman, G. D., Sloman, K. A. & Wilson, R. W. Does structural enrichment for toxicology studies improve zebrafish welfare? *Applied Animal Behaviour Science* **139**, 143–150 (2012).
59. Franks, B. What do animals want? *Animal Welfare* **28**, 1–10 (2019).

60. Špinka, M. Animal agency, animal awareness and animal welfare. *Animal Welfare* **28**, 11–20 (2019).
61. Kittilsen, S. Functional aspects of emotions in fish. *Behavioural Processes* **100**, 153–9 (2013).
62. Egan, R. J. et al. Understanding behavioral and physiological phenotypes of stress and anxiety in zebrafish. *Behavioural Brain Research* **205**, 38–44 (2009).
63. Nguyen, M., Stewart, A. M. & Kalueff, A. V. Aquatic blues: Modeling depression and antidepressant action in zebrafish. *Progress in Neuro-Psychopharmacology & Biological Psychiatry* **55**, 26–39 (2014).
64. Wright, D., Ward, A. J. W., Croft, D. P. & Krause, J. Social organization, grouping, and domestication in fish. *Zebrafish* **3**, 141 155 (2006).
65. Robison, B. D. & Rowland, W. A potential model system for studying the genetics of domestication: Behavioral variation among wild and domesticated strains of zebra danio (*Danio rerio*). *Canadian Journal of Fisheries Aquatic Sciences* **62**, 2046–2054 (2005).
66. Séguret, A., Collignon, B. & Halloy, J. Strain differences in the collective behaviour of zebrafish (*Danio rerio*) in heterogeneous environment. *Royal Society Open Science* **3**(10) (2016).
67. Wright, D., Rimmer, L. B., Pritchard, V. L., Krause, J. & Butlin, R. K. Inter and intra-population variation in shoaling and boldness in the zebrafish (*Danio rerio*). *Naturwissenschaften* **90**, 374–377 (2003).
68. Williams, T. D., Readman, G. D. & Owen, S. F. Key issues concerning environmental enrichment for laboratory-held fish species. *Laboratory Animals* **43**, 107–120 (2009).
69. Piato, Â. L. et al. Unpredictable chronic stress model in zebrafish (*Danio rerio*): Behavioral and physiological responses. *Progress Neuro-Psychopharmacology Biological Psychiatry* **35**, 561–567 (2011).
70. Brydges, N. M., Boulcott, P., Ellis, T. & Braithwaite, V. A. Quantifying stress responses induced by different handling methods in three species of fish. *Applied Animal Behaviour Science* **1**, 295–301 (2009).
71. Galhardo, L., Vital, J. & Oliveira, R. F. The role of predictability in the stress response of a cichlid fish. *Physiology Behavior.* **102**, 367–372 (2011).
72. Bassett, L. & Buchanan-Smith, H. M. Effects of predictability on the welfare of captive animals. *Applied Animal Behaviour Science* **102**, 223–245 (2007).
73. Wang, N. Increasing the reliability and reproducibility of aquatic ecotoxicology: Learn lessons from aquaculture research. *Ecotoxicology and Environmental Safety* **161** (2018).
74. Sloman, K. A., Baker, D. W., Wood, C. M. & McDonald, G. Social interactions affect physiological consequences of sublethal copper exposure in rainbow trout, *Oncorhynchus mykiss*. *Environmental Toxicology and Chemistry* **21**, 1255 (2002).
75. Garner, J. P. The Significance of meaning: Why do over 90% of behavioral neuroscience results fail to translate to humans, and what can we do to fix it? *ILAR Journal* **55**, 438–456 (2014).
76. Richter, S. H., Garner, J. P. & Würbel, H. Environmental standardization: Cure or cause of poor reproducibility in animal experiments? *Nature Methods* **6**, 257–261 (2009).

8 The Mouse

Brianna N. Gaskill and Kelly Gouveia

CONTENTS

THE ANIMAL

It is important to realize that not all types of laboratory mice are similar. There are clear behavioral differences between genetic inbred or outbred mice (see Jackson Laboratory's Phenome Database). There are even documented genetic and behavioral differences between substrains of mice (C57BL/6J vs. C57BL/6N).[1] Thus, understanding and reporting specifically which types of mice are used is extremely important in scientific writing,[2] but equally important for animal-centric management. For example, it is important to understand which types of mice are highly aggressive (e.g., SJL and FVB). This knowledge should alter the vigilance of care staff for wounding and perhaps consider the option of housing males singly.

In general, domesticated laboratory mice express many of the behaviors observed in wild *Mus musculus*. However, there are likely differences in terms of reduced stress and fear responses that are characteristic of many domesticated species.[3] Although similar, it is crucial for true animal-centric management to understand the specific behavioral drives of laboratory mice. Several excellent mouse behavior resources currently exist[4–7] and we would highly recommend anyone working closely with laboratory mice to become familiar with them.

To really understand mice, the laboratory animal science community needs to understand what behaviors helped mice to survive extremely varied and harsh climates as well as colonize every continent except Antarctica.[4,8] The telos, or behavioral characteristics, that define rodents are that they are highly motivated to chew, highly reproductive, are prey for many different types of predators, and predominantly nocturnal.[4] These unique behavioral characteristics can create housing challenges because mice will attempt to chew through their enclosure if given the opportunity, have been known to mate through cage lids if they escape, likely perceive humans as predators, and find the brightly lit rooms for human vision aversive.[9]

Despite being housed in less than desirable conditions, if mice are allowed to use the behaviors that helped them cope in harsh wild climates, they too can thrive in the laboratory. Control, or the perception of control, is the single best method for reducing the impact of stress in any environment.[10] Behavior specifically allows these small mammals to exert control over their environment. Behavioral control not only is beneficial to the mice but can be a visual signal to caretakers, indicating their overall well-being.

Nest building is a perfect example of a behavior that allows mice control over a stressor, such as typical laboratory temperatures,[11] while also giving an indication of their well-being.[12–14] Nest building can be considered a thermoregulatory behavior that is increased at colder temperatures and decreased at warmer ones.[15–18] Thus, it is unsurprising that when mice are provided with appropriate material, most build fully covered, dome-like, nests in the laboratory.[19–21] Large, fully enclosed, nests can indicate that mice, even in groups, may be cold. The shape of the nest, as well as the behaviors that create the final product, can be clues to staff about how the mice are faring. Nest-building behaviors can be useful because reproductive and non-reproductive mice are highly motivated to gather materials to build a nest.[19,22] Thus, if mice are no longer building a nest (i.e., material may be scattered all over the cage or simply laying on top of the material), something may be going on inside the cage that requires more detailed observation. For instance, when mice are ill or in pain, nest building is reduced, resulting in more open, or low walled, nests.[12,14,23] Furthermore, the level of aggression, or wounding, in the cage has been found to correlate with lower walled or scattered nesting material.[24] Thus, nest shape is a non-specific behavioral indicator that can be used to identify general changes in welfare at the cage level.

Despite being phenomenal nest builders, laboratory mice also burrow. This behavior can also be used to identify changes in animal welfare, disease onset, and brain dysfunction. Deacon et al.[25] developed a burrowing protocol that is sensitive enough to identify progression of disease in Alzheimer's models,[26] prion disease,[27,28] hippocampal lesions,[29] as well as strain-specific burrowing differences.[30] While useful scientifically, the behavior can also be beneficial in assessing mouse well-being. Jirkof et al.[31] showed that burrowing behavior was delayed in mice that received a laparotomy surgery without analgesia. Surgical controls however, those that just received anesthesia and analgesia drugs without the surgery, were also delayed in burrowing compared to baseline measurements. Thus, burrowing behavior is sensitive to pain as well as the effects of the drugs themselves and should be carefully evaluated to determine if animals are still in pain post-surgery. Jirkof et al.[32] also showed that burrowing behavior could be used to identify inflammation in a dextran sulfate sodium induced colitis model and perhaps be utilized as a humane endpoint to reduce suffering. Thus, burrowing behavior, similar to nesting, is sensitive to welfare changes but not specific enough to identify exactly what the problem may be. Therefore, in practice, these behaviors can be used to identify cages that require more detailed observation to determine the specific cause of the welfare change.

While the absence of highly motivated behaviors, such as nesting or burrowing, can indicate a change in animal welfare, observing the occurrence of others can also be a clue to how animals are coping in the laboratory. Abnormal behaviors are defined from an animal's perspective as quantitatively or qualitatively unusual.[6] Our current understanding of why abnormal behaviors occur and what they mean is reviewed in Chapter 4 of this book but here we will focus on specific behaviors common in laboratory mice. The most commonly observed abnormal behaviors in mice are stereotypies or abnormal repetitive behaviors. The defining features of a stereotypy are that it is repetitive, unvarying, and seemingly functionless.[33] Particularly in mice, the behavior stems from a mouse's motivation to escape the home cage.[34] Mouse stereotypies come in many shapes and forms such as bar-mouthing, jumping, route-tracing, and flipping.[33] Stereotypies are considered a malfunctional type of abnormal behavior because animals have a similar dysfunction in inhibitory brain regions as humans who display stereotypies.[35] Thus, the brains of these animals are not normal and the behavior cannot be completely cured once established. Because these behaviors cannot be "cured," stereotypies are often referred to as behavioral scars because they persist even after animals are put in ideal environments. Stereotypies generally develop around sexual maturity and once established, environmental enrichment will not be as effective in reducing them.[33] If enrichment is provided at an early age however, it can help reduce the likelihood of these behaviors developing.[36]

Barbering and ulcerative dermatitis are two other types of behaviors, from the malfunctional category of abnormal behaviors, that are detrimental to a mouse's welfare. Barbering is where the fur or whiskers are plucked from a mouse by the individual or his/her cage mate. It is most commonly seen in C57BL/6 mice, females, and around hormonal changes, such as sexual maturity.[37] It is not a dominance-related behavior and in fact, both dominants and subordinates will engage in the behavior.[37] The loss of fur or whiskers, on the surface, does not seem to be terribly impactful on a mouse's welfare, however the loss of whiskers eliminates a primary mode mice use to explore their environment[4] and hair loss increases heat loss (from the lack of fur cover), metabolism, and oxidative stress.[38] However, barbering can be prevented and treated by supplementing N-Acetyl Cysteine in a mouse's diet.[38]

Ulcerative dermatitis is a condition characterized by lesions found at the base of a mouse's neck, near the shoulder blades, that are the result of excessive, and abnormal, scratching behavior.[39,40] These bloody, self-inflicted open wounds may require euthanasia. This particular abnormal behavior appears to have a similar etiology to skin picking behavior in humans.[40] Ulcerative dermatitis accounts for 22% of clinical cases at a major research institution[39] and often leads to premature euthanasia. Interestingly, if mice are already barbers, they seem to be less likely to develop ulcerative dermatitis.[40] However, if mice do develop it, a simple toenail trim resolves 93% of animal lesions within 14 days.[41] An open-sourced instructional video demonstrating how to trim mouse nails accompanies the manuscript in the supporting information by Adams et al.[41] This simple and quick technique has the potential to reverse one of the most serious welfare problems in laboratory mice.

Another abnormal behavior category is maladaptive behavior. Unlike malfunctional, animals are considered normal and the behaviors they do are natural, but they occur much more often than they would in the wild. One mouse behavior often considered a maladaptive behavior is aggression. It is normal and is used to protect offspring and territories in the wild but happens at an abnormal level in the laboratory. In wild populations, aggression often stops when an intruder exits the territory or line of sight is broken.[5] However, it is near impossible for these aspects to end aggression in a typical, minorly enriched, shoebox mouse cage. Aggression is a major welfare issue in laboratory mice and the second most common reason for veterinary clinical care.[39] Injury associated with aggression is often extreme, resulting in large open wounds. High levels of aggression destabilize dominance hierarchies,[42] which can, in turn, induce immune suppression,[43] resulting in poor welfare and unwanted variation in scientific models.

Infanticide, or pup cannibalism, is also considered a maladaptive behavior. Wild mice are known to kill their pups when a new male takes over a territory (see review by Latham and Mason[4]). In the laboratory, it is believed that high levels of infanticide in a breeding colony may be due to odors from other male mice housed within the same room or from human odors, deposited after handling a new litter. This is why in many breeding colonies, litters are not touched for several days, up to a week, after birth. However recent literature indicates that infanticide may not be the welfare issue we think it is. Weber et. al.[44] found no evidence of infanticide in female C57BL/6 mice. Pups were indeed observed to be eaten, however this only occurred after they had stopped moving. In further work,[45] females who spent more time nest building prior to parturition, were more likely to have their litters survive. Interestingly, females that lost their litters engaged in more parturition-related behaviors, perhaps from prolonged labor. Thus, the high rate of litter loss seen in several inbred strains[46] is likely due to issues with parturition and perhaps not a maladaptive abnormal behavior at all.

COGNITIVE ABILITIES

How intelligent are mice? We often underestimate the cognitive abilities of these small mammals because we ask them questions in ways that make sense to humans, not mice. When designing cognitive tasks, it is important to remember that mice do not experience the world in the same way we do.[47]

Indeed, humans are visually driven while mice rely more heavily on olfaction, auditory, and tactile senses but have comparatively poor eyesight.[4,47] Understanding these sensory differences is essential as the nature of the cues provided in behavioral paradigms can have a substantial impact on the performance of the animal.[47–49] It has often been reported that behavioral data from mice can be variable.[50] However, this variability is likely due to environmental influences and stressors[51] or using inappropriate stimuli for a particular species.[48] Unfortunately, visual and food stimuli are still the predominant cues used in cognitive testing, with food deprivation still widely used despite the potential for compromising welfare. Olfactory cues, on the other hand, have been successfully introduced in a number of cognitive tasks, as is the case in a modified open field test, place preference, and habituation tasks.[52,53] For example, darcin, a major urinary protein from male mice that attracts females, has been strategically applied to test spatial memory in mice.[54] Other sensory modalities have been applied to more complex cognitive tests, such as interdimensional extra-dimensional set shifting, a test with clinical relevance to disorders such as autism, obsessive-compulsive disorder, schizophrenia, and Tourette's syndrome.[55] The test was altered[55] to incorporate tactile and olfactory stimuli. Mice were able to accomplish the test, originally thought to be impossible for this species, and showed internal construct and predictive validity. In addition, the timing of when mice are tested (light vs. dark phase) should also be considered since it is influential on their performance.[56,57] In particular, mice are nocturnal and therefore are more likely to investigate test stimuli when tasks are conducted in the dark phase. Testing during the light phase is unfortunately still common practice because of human convenience.[58] Thus, more often than not we are asking mice questions in a way that makes sense to us, as humans, or when convenient with our diurnal cycle. Perhaps instead of asking how smart mice are, we should be considering if we are asking them to learn or respond in a species-specific way.

EMOTIONS

Rodent affect, or their emotions, has been studied primarily in rats due to their unique vocalizations.[59] However, the emotionality of mice is a topic rarely discussed in laboratory animal science. The reason for this is unknown but it begs the question, why would we not expect mice to live similar emotional lives to rats? Although animal affect is more broadly covered in Chapter 3, we believe that to truly embrace an animal-centric view of how we house, use, and care for mice, it is important to acknowledge that they too live emotional lives. This is especially important since 1) millions of mice are used for scientific research every year and the emotional welfare of so many animals is vitally important and, 2) if we are using them to model human emotional circuitry, it is difficult to deny that mice experience similar emotions. In fact, a recent publication illustrated that like humans and primates, facial expressions in mice appear to be real-time reflections of their affective state, which can be elicited when appropriate regions of the brain are stimulated.[60] While we may not fully understand the extent of their emotional competence, similarities are evident given the widespread reliance on mice as models for human mental health.[61] It is also hard to deny that basic emotions give animals an advantage evolutionarily since they inherently help to anticipate future events.[61] Comparatively in humans, the most powerful affective experiences stimulate deep subcortical areas of the brain.[61] Primal feelings, similar in all mammals, originate in these subcortical structures.[61] Jaak Panksepp, well known for his work in affective neuroscience and "rat tickling," describes seven basic emotional systems (see review[61]) that we will use here to investigate what we know about mouse affect.

The SEEKING system is an appetitive motivational system that animals use to find what they need for survival.[61,62] In scientific studies where animals are given control of turning on and off these circuits, they clearly indicate what they like or dislike.[63] Essentially, these circuits indicate the reward of an item or substance. Mice have increasingly been used to study reward, including in knockout and transgenic mice.[63,64]

Of all the emotional circuits, the existence of the FEAR system is likely the least debated since fear testing is prominently used in experimental psychology. We see similar physiological responses in mice as we do in humans such as increased respiration, heart rate, and release of catechol-amines.[9,65] FEAR is an important survival mechanism that helps animals avoid the possibility of experiencing pain.[62]

Emotions involved in the RAGE system induce aggressive behaviors when animals are irritated or restrained but also helps animals to defend themselves by arousing fear in their opponents.[62] Activation of these emotions often leads to attack, biting, and fighting.[61] RAGE is, unsurprisingly, a negative emotion. When given the choice, rats will turn off stimulation of brain areas that evoke these emotions[66] as well as avoid areas where stimulation previously occurred.[61] Research on mouse aggression is predominantly focused on territorial aggression, using the resident-intruder tests.[67] It is possible that this type of aggression may stimulate the RAGE circuit, since animals may feel that they need to defend themselves.[67] However the aggression observed in the home-cage is unlikely to be territorial and instead related to dominance since mice must create social relationships within the cage.[68] Whether general home-cage aggression triggers the RAGE circuit is unknown.

An animal's general drive to reproduce, which is regulated by sex hormones and other neuro-peptides, involves the LUST system.[61] Mice are a highly fecund species, reproducing at high rates when resources are plentiful. In fact, it is a defining characteristic of mice. Although we consider this particular emotion being strongly driven by hormones, emotional circuits are also involved. A review by Pfaus et al.,[69] describes rodent behaviors, primarily in rats, that are analogous to human sexual arousal, desire, reward, and inhibition. However, the question of whether mice experience heightened LUST emotions is a possibility because of some similarities to human arousal[70] and the fact that these animals are so driven to reproduce.

The CARE system is responsible for assuring that parents, typically the female, care for the off-spring.[62] This particular system was initially studied in a dam's response to vocalizations from her pups. Again, this emotion makes evolutionary sense, due to the extensive parental investment mice make when rearing offspring. It is advantageous for dams to be sensitive to their offspring's affec-tive states to ensure their likelihood of survival.[71]

The PANIC/GRIEF system was simultaneously studied with the CARE system but in the pups instead of the dams. Mouse pups give off ultrasonic distress calls when they are separated from the nest and conspecifics.[72] These isolation calls were the first scientific evaluation of affective states in animals and their vocalizations.[72] Although primarily studied in offspring, this system promotes sadness and depression in adults.[62]

The last emotional system is PLAY, or "physical social engagement," where animals learn the affective value of social interactions.[62] PLAY is associated with positive emotional states and is reduced when animals have poor welfare or when vital necessities are not being met.[73,74] Mice have a more primitive type of social play than rats do and instead engage primarily in solitary locomotor play.[75] This is characterized as "popcorn"-like jumping behaviors with chasing and darting away from one another.[76] While usually categorized as locomotor play, the jumping can be contagious and synchronized darting may be observed.[76] Mice also engage in play fighting; however, this is only seen in an extremely small percentage of all mouse play.[75] Play is reduced in pre-weaning mice when space is limited or in crowded situations.[77] Even within genetically identical mice, individual differences in levels of play have been documented.[78]

THE ENVIRONMENT

HOW TO ENRICH

The provision of environmental enrichment has increased in scientific importance and popularity in recent years. However, we are often looking for a one-size-fits-all answer to what we should provide as environmental enrichment. We, visual humans, are generally more concerned with color and design and not about the functionality of an item to the mouse. In order to embrace mouse-centric management, we need to be thinking about what specific stressors the individual mouse or a particular model may be experiencing, and then provide enrichments that best allow that animal to control that stressor. Würbel and Garner[79] describe a true enrichment as being biologically relevant, meaning that they address or provide an outlet for a species' particular behavioral need. Most behaviors exist to help an animal reach an end goal. Behaviors such as searching for food (foraging) are called appetitive behaviors and the goal behavior (in this case eating the food) is the consummatory behavior. Appetitive behaviors, such as foraging, are often neglected because we present the animal with food in a sterile dish, requiring no work or foraging. The act of searching or foraging is an internal drive that in the wild must precede feeding. Taking these behaviors out of their order of operation can be jarring to an animal over time, and in fact can lead to abnormal behavior development (see Chapter 4 for more details). Allowing animals to engage in these appetitive behaviors may be as mentally rewarding as the goal itself. An enrichment allows an animal some semblance of control, facilitating homeostatic goals of behavior, and will shield that animal from a particular stressor. This ability to control or predict stressors reduces the physiological impact of stress on the body.[80] While the most important, this last characteristic is perhaps the most difficult to implement in practice.

In an ideal world, we would provide environments similar to those provided to pet mice. They are often much more complex, diverse, and challenging than traditional laboratory environments. However, some very complex and enriched housing has been implemented in real vivariums, which actually advanced our understanding of, and interaction between, physiological systems. For example, Cao et al.[81,82] compared tumor growth between mice housed in normal shoebox cages and large enriched enclosures (Figure 8.1). The enriched housing stimulated the hypothalamus to produce brain-derived neurotrophic factor (BDNF), which activated the hypothalamic-sympathoneural-adipocyte axis. This sympathetic nervous system pathway suppressed tumor growth. If the authors had not pursued the observation that enriched environments upregulated BDNF expression,[83] we would not understand how sensory, cognitive, social, and motor stimulation can influence tumor growth. Perhaps this will provide vital information about risk factors in human environments. It is exciting to think that these impractically enriched enclosures enhanced animal welfare while improving our understanding of basic biology and the data's translation to human medicine.

FIGURE 8.1 This extensively enriched enclosure was used to investigate the effects of environment on tumor growth. Male mice (18–20 per cage) were successfully housed in 1.5 m × 1.5 m × 1.0 m enclosures that included running wheels, tunnels, igloos, huts, retreats, wood toys, a maze, and nesting material in addition to food and water. The image is reprinted with permission from Elsevier.[80]

PRACTICAL ENRICHMENTS

The practicality of enrichments is often a limiting aspect to the improvement of mouse environments. Factors that are considered when deciding which enrichments to provide for vast numbers of mice are the purchasing cost, implementation time and ease, and how well it holds up after cleaning. Obviously, none of these factors are animal-centric and likely lead to easier implemented items, instead of effective ones. Implementation for even the most scientifically proven, effective, practical, and cost-effective enrichment, nesting material, can still be an uphill battle in the United States. The benefits of nesting material for mouse behavior, welfare, and scientific outcomes[19–21] are undeniable and some argue that it should be considered a basic need instead of an enrichment.[84] Although the provision is not hotly debated, providing the recommended minimum amount of nesting material (8–10 g),[19] is often met with reluctance. When mice are provided more than 6 g of material,[19] they are able to build such elaborate, dome-shaped, nests that they cannot be easily visualized during routine daytime wellness checks (Figure 8.2).[84] This often leads to mice being provided small amounts of material (less than 6 g)[85] that do not allow them to build sufficiently insulating nests to reduce cold stress.[19] Thus, the practical implementation of nesting material (providing a small amount that limits the behavioral goal) may not fully meet the characteristics of an enrichment.[79]

Similar to nesting material, social housing is still often considered an enrichment, even though group housing is the default paradigm for social species in many countries.[86,87] Although one of the five categories of enrichment outlined by Young,[88] it is crucial to the proper development of social species due to the varied mental stimulation it provides.[88,89] In terms of practicality, social housing, on the surface, appears to be the most practical enrichment. Housing multiple mice per cage reduces the total number of cages and resources needed for a study. However, mice can be a challenging species to house socially. Females generally cohabitate peacefully but males can rip each other apart within hours. Thus, with male mice we have to consider the welfare trade-offs between the stress of single housing[90–92] or the potential pain, suffering, and experimental variability resulting from aggressive interactions.[43,93] Thus, excessive aggression challenges the practicality of social housing as an enrichment for male mice. Research on how to reduce aggression between male mice as well as what might trigger the behavior has produced varied results.[68] However, one easy strategy, the transfer of nesting material at cage change, appears to have consistent results.[94]

Another practically implemented enrichment is a physical shelter. These items are easily implemented into standard shoebox mouse cages, come in several colors, and are easily cleaned. Physical shelters can include polycarbonate igloos, paper pulp domes, or even tubes or tunnels. Shelters create a retreat space which has been shown to reduce fear and abnormal behavior development,

FIGURE 8.2 These photos illustrate how thick and large mouse nests can be when given sufficient material to build with (10–12 g). An albino mouse is within the nest shown on the right but the animal cannot be visualized. The photo on the left shows a mouse sleeping inside another fully enclosed nest but a small window into the central cavity is visible through the side of the cage (Photo credit: B. Gaskill 2019).

and increase comfort behaviors across various captive species (see review).[95] Ultimately, retreat spaces allow animals to cope with, or have control over, a potential threat.[95] Laboratory mice prefer access to nesting material over a physical shelter[96] but if it is the only option, they are likely to value them. Wild house mice often establish their home territories around nest boxes[5] but physical structures (such as igloos) can instigate territorial fighting in the laboratory.[42] The fact that they are willing to set up a territory and fight for this resource, demonstrates its value to the mice. Tubes on the other hand, like those used for tunnel handling (described below), have not been found to increase aggression.[97] However, due to this potential aggression, we caution the provision of physical structures in cages of healthy male mice because they may trigger natural but undesirable behaviors.

Healthy mice readily engage in nest building behaviors but if they experience any kind of malaise or discomfort they stop building.[24,98,99] In these circumstances, a physical shelter is superior for immediately providing a dark, safe, space for mouse models of disease that may experience malaise. However, if mice are healthy and motivated to build a shelter (i.e., nest), why should we deny them the opportunity to engage in rewarding, goal-directed, behaviors by instead giving them a physical shelter? Therefore, physical shelters are a practical and effective enrichment in certain circumstances.

HANDLING

HANDLING FROM THE ANIMAL'S PERSPECTIVE

In the laboratory, it is inevitable that research animals come into contact regularly with human handlers, either for routine husbandry procedures or experimental purposes. Human-mouse interactions are therefore an essential part of a laboratory animal's social environment and have the potential to enrich the lives of both the animal care staff as well as the animals. The first and most important point that the handler should remember when approaching mice is that they are a prey species, meaning that they will naturally avoid being caught when humans attempt to handle them.[100] In contrast to other small laboratory mammals, such as rats, guinea pigs, and gerbils, which are thought to have a more docile nature and are easier to tame, mice are often reported to bite.[101]

A species-specific point to consider is that mice are usually approached from behind and above, which is how they are caught by predators in the wild, and could certainly justify why they are often reported to bite.[102] Yet, approaching the animal from above is inevitable in the laboratory. However, approaching the animal from behind can easily be avoided, as discussed later in this chapter, using alternative approaches that can effectively reduce handling-related anxiety, contributes to improved mouse welfare.

Indeed, techniques for handling any laboratory species should be carefully considered for both welfare and safety reasons. It is particularly essential to consider methods that are effective and that ideally allow for the animal to express species-specific behavior when handled.[103] This allows for the animal to exert some degree of control over the interaction with the handler, and subsequently dictates how the animal will anticipate and react to future interactions. Thus, if the animal has a positive experience, then anxiety associated with handling will subside. However, if it is an aversive experience, the interaction will reinforce a negative association to humans and handling.[104] In this chapter, we present practical tips on human-mouse interactions that minimize handling stress, thus improving welfare and practicality for the handler.

HANDLING FROM THE HANDLER'S PERSPECTIVE

Before discussing best practices for handling laboratory mice, it is important to first consider the human-mouse relationship and how this is likely to differ from other laboratory species. This is important to establish a clear understanding of the practical needs and constraints involved in the care of this particular species in the modern laboratory. In modern facilities, animal care staff are likely to handle numbers of mice that reach the order of hundreds per day. Yet, historically mice have been classified as one of the least preferred species, in contrast to rats and other larger charismatic mammals, such as cats and dogs.[105] It is possible that this could be due to several factors. In particular, the vast numbers of mice kept in research facilities impose substantial practical constraints on animal care staff for individual animal handling, which inevitably limits contact time with the handler to brief seconds during routine husbandry procedures. Similarly, "bulk" handling of large numbers of animals, also makes recognizing individuals considerably more difficult, and thus collectively many contribute to poorly established, individual, animal-human bonds in the laboratory.

Moreover, positive human-animal interactions are more likely to occur in species that we readily interpret facial expressions and have a greater understanding of specific behavioral needs (e.g., dogs and cats). We know comparatively less about interpreting mouse facial signals, behavioral needs, or specific needs for appropriate handling.[103] In particular, the need to seek shelter is an obvious evolutionary trait that these animals have conserved from their wild counterparts.[106] Yet, it is also important to take this into account during handling, as a greater understanding of a mouse's behavioral responses is likely to be a more positive mouse-human interaction in the laboratory, and could impact job performance and satisfaction of laboratory animal care staff. Indeed, the rewarding properties and reinforcing ability of positive human-animal interactions are known to be one of the leading motivators that sustain a long-term commitment and career in this industry.[105] There is also a direct correlation between positive human-animal interactions in the laboratory and the development of a greater understanding and empathy for the research species,[107] thus contributing to the culture of care, discussed in Chapter 2.

REFINING MOUSE HANDLING IN THE LABORATORY

The standard practice for handling mice is to grasp the animal by the tail. This involves lifting the animal by the base of the tail, with body weight supported on the back of the hand or suspending the animal completely by the tail for short transfers.[108] It is possible that this method has become standard for practical reasons, in particular because it is relatively simple to carry out and possibly due to its historic widespread recommendation in the literature.[109,110] Yet, despite the common practice of tail handling, there is no scientific validation supporting its use. Further, tail handling increases the risk of being bitten, due to it being inherently a predatory approach for this species, which ultimately renders this method of little practical benefit to the handler. Yet, a growing body of research has emerged in the last decade showing that two particular alternative techniques have the potential to substantially improve animal welfare and scientific outcomes.[111]

These consist of using a tunnel and allowing the animal to "walk" in or cupping the animal onto open hands.[102] These methods are thought to reduce the animal's stress response associated with the handling procedure. Tunnel use, and the subsequent reduction of handling stress, has also proven beneficial in reducing experimental variation in cognitive[52] and pharmacological studies.[112] Reducing the mouse's emotional reactivity to handling most likely minimizes physiological and behavioral changes associated with stress during these tasks, which could mask the true experimental effect. Other areas where these methods are beneficial in reducing experimental variation are in metabolic studies, where non-aversive handling reduces corticosterone (a stress hormone) and glucose levels in the blood.[113,114] Thus, use of non-aversive methods has clear benefits for improving mouse welfare and experimental research. Although the science provides evidence for its effectiveness, only 18% of survey respondents indicated the sole use of non-aversive methods when handling mice.[115] However, the majority of respondents (43%) reported the use of both non-aversive and tail handling methods.[115]

Practical Considerations for Non-Aversive Handling

Non-aversive handling is relatively simple to carry out once the handler is practiced and confident with the methods involved. For the tunnel method, the handler should prefer to use the animal's own home cage tunnel over an external tunnel, so that it carries the animal's and its cage mate's familiar scent. This is recommended as it is possible that contact with unfamiliar scents from animals housed in other home cages could potentially cause inter-male aggression, and it has been suggested that this could lead to potential physiological changes to male or female reproductive status,[103] thus introducing experimental variation. When approaching the animal, the handler should use one hand to hold the tunnel in the home cage, and use their free hand to guide the animal towards the tunnel entrance (Figure 8.3).[103] This method is relatively simple to carry out even for inexperienced handlers, and therefore offers a suitable alternative to tail handling. There are a variety of tunnel choices available in terms of colors and materials (Figure 8.4), but an autoclavable plastic material (e.g., acrylic) tends to be best due to the slippery surface, which allows the handler remove the animal easily. In terms of size, generally a diameter of 50 mm wide and 12–18 cm long is recommended.[102,103] The color of the tunnel does not seem to affect the animal's response to handling, but choosing a transparent one may be easiest for inspecting the animal. Although time is often a concern surrounding the implementation of this technique, those currently using it state that they use it is because of its speed and efficiency.[115]

For cup handling, the animal should be scooped between open hands so that s/he is free to sit or walk on the palms.[102] Generally, the mouse is first habituated to being in the hand prior to carrying

FIGURE 8.3   Tunnel handling. The handler uses one hand to hold the tunnel in a fixed position in the cage, while their free hand is used to guide the animal towards the tunnel entrance.

FIGURE 8.4   Tunnels made of different materials and tinting that can be used for mouse handling.

out this method, to avoid him/her leaping from the open hands. This can be achieved by picking up the animal in a tunnel, then tipping him/her out onto an open hand, where s/he can be held (up to 20 s). The other option is to confine the animal between closed hands (up to 20 s) on the initial attempts at cupping (Figure 8.5). [102] Here, it is worth noting that the animal will likely press his/her nose against the closed hands in an attempt to escape, and depending on the strain (e.g., BALB/c) the animal may even nibble the hand. This is part of the animal's normal approach-avoidance conflict[116] associated with the novelty of being enclosed in hands, which should rapidly subside once the animal is habituated to the technique. These methods tend to be most helpful, in particular, for jumpy strains (e.g., C57BL/6) or juvenile animals (four to six weeks old), as more docile strains (e.g., ICR-CD1 mice) and older animals (over eight weeks old) may be more accepting of open handed cupping, without habituation. The main disadvantage of this method, relative to tunnel handling, is the initial time investment needed for habituating the animal to the hand. Yet, cup handling offers a unique opportunity for the handler to develop a greater bond with the animal. The handler is likely to observe that the animal will start to spend more time investigating the hand (e.g., sniffing and walking up and down the hand), which can also increase the handler's confidence and interest/ empathy for this particular species. Moreover, cupping often resulted in mice voluntarily walking up onto the hand directly from the home cage (personal observation by K. Gouveia). Thus, the level of tameness observed here shows how it is possible for the handler to create a positive relationship with this particular species, as laboratory mice in general seem to have the capacity to behave as "pets" when appropriate habituation and training are considered.

Further, non-aversive handling can be used for animals undergoing minor husbandry procedures in the laboratory. For example, when an injection or blood draw needs to be done, mice are usually restrained by scruffing. When scruffed, the mouse is held firmly for the procedure by grasping the skin at the back of his/her neck between the handler's thumb and forefingers. Non-aversive handling methods can be combined with scruffing by picking up the animal from his/her home cage by cupping or using a tunnel, then placing him/her on a grip surface for scruffing. In contrast to tail handling, these methods are an effective refinement to handling related stress in mice undergoing scruff restraint.[102,117] This is because tameness towards the handler and the anxiety-reducing effects of handling are still sustained even when mice undergo repeated scruffing,[117] thus demonstrating the crucial role of handling methods in controlling stress.[102,103] Further guidance on non-aversive handling techniques and practical tips for incorporating best handling practice into routine laboratory procedures can be found at: https://www.nc3rs.org.uk/mouse-handling-video-tutorial.

FIGURE 8.5 In order to habituate mice to cupping handling, a mouse is held gently between closed hands for several seconds.

STANDARDIZING NON-AVERSIVE HANDLING IN RESEARCH

For research facilities considering the uptake of non-aversive methods as standard practice for handling laboratory mice, there are likely to be substantial benefits to animal welfare, the human-animal bond, as well potential improvements to experimental reliability of mouse studies. Yet, it is essential to consider the benefits of refining handling practices against potential practical limitations for a more robust implementation within and among facilities. The first and most important point to consider is that a single handling method may not necessarily suit the handling requirements of all on-going research at a given facility. There may be areas of research as explored previously (e.g., cognitive studies) where non-aversive handling could greatly benefit animal welfare and experimental reliability, but there are likely to be others where taming effects of handling could be limited or undesirable from a 3Rs perspective. Limitations to handling could occur, in particular where mouse models of stress are created to study effects of psychoactive drugs in pharmacological studies, whereby taming the animal could minimize behavioral and physiological responses. Indeed, a recent study by Clarkson et al.[112] shows that tail handling is likely to increase behaviors indicative of depression, in particular anhedonia (often measured as reduced sucrose consumption), in mouse models of psychiatric disorders, in contrast to using a home tunnel. Thus, the individual aims of the research need to be carefully considered against potential effects of handling on welfare and experimental variation for effective decision making of whether to implement handling refinements.

FURTHER CONSIDERATIONS ON LABORATORY MOUSE WELFARE: WHERE DO WE GO FROM HERE?

A growing body of scientific research has emerged since Russell and Burch published their principles for humane experimentation in 1959. This increase in interest and publication demonstrates the benefits of refining housing and husbandry procedures for improving laboratory animal welfare and scientific outcomes. Yet, intense debate has arisen, particularly over the past two decades, weighing the benefits of enrichment for animals against the potential negative scientific outcomes such as increased experimental variation; however, data is conflicting. This conflicting data is possibly due to a combination of factors, such as the level of complexity the particular environment provides, its species-specific appropriateness, the extreme standardization in animal-based research that leads to unreproducible results,[118] the different types of mice used, and the variety of outcome measures used to assess variability (strong bias towards physiological measures).[119] Moreover, recent research has shown that even simple enrichments, such as nesting material and shelters, do not compromise variability in an array of parameters, from hematology, biochemistry, energy metabolism, pain perception, and pathology in laboratory mice.[119] Therefore, given the potential for biologically meaningful enrichments to substantially improve an animal's ability to exert control over his/her environment, and thus contribute towards a greater quality of life, it seems justifiable from an ethical perspective that we provide mice, at a minimum, species-specific enrichments in the form of social housing and increased cage complexity with nesting material and shelters.

Further, the extent to which the laboratory animal science community can move beyond simple enrichment toward idealized housing conditions, i.e., the so called "super-enriched" enclosures (e.g., "Disneyland for mice" – created in the Cao et al. study[83]), is possibly still an unrealistic expectation for large-scale implementation. However, this may not even be paramount for substantial improvements in welfare, as for example, floor space is considered to be less important in terms of providing opportunities for mice to exercise species-specific behaviors (e.g., climbing) in comparison to vertical space in the enclosure. Further, it is possible that manufacturer-led research on cage design and complexity, from a mouse's perspective, could substantially aid in devising practical features that would enable enhancement and practical implementation of increased cage complexity on a large scale. An example of this is the double decker cage, which is now widely available for housing rats and has been validated for collecting cardiovascular data via telemetry while socially housed.[120] It is possible that similar research-informed alternatives could be devised for mice, particularly considering species-specific needs and manageability. In the meantime, careful selection of enrichments that meet species-specific behavioral needs, and that empower the animal with a level of control over his/her environment, are still likely to benefit the welfare of the many millions of mice kept in laboratories worldwide.

REFERENCES

1. Simon, M. M. et al. A comparative phenotypic and genomic analysis of C57BL/6J and C57BL/6N mouse strains. *Genome Biology* **14**, R82 (2013). doi: 10.1186/gb-2013-14-7-r82
2. Kelmenson, P. There is no such thing as a C57BL/6 mouse. *Jax blog* Vol. 2019 (2016). https://www.jax .org/news-and-insights/jax-blog/2016/june/there-is-no-such-thing-as-a-b6-mouse#. Accessed 2019.
3. Trut, L., Oskina, I. & Kharlamova, A. Animal evolution during domestication: The domesticated fox as a model. *BioEssays* **31**, 349–360 (2009). doi: 10.1002/bies.200800070
4. Latham, N. & Mason, G. From house mouse to mouse house: The behavioural biology of free-living *Mus musculus* and its implications in the laboratory. *Applied Animal Behaviour Science* **86**, 261–289 (2004).
5. Crowcroft, P. *Mice All Over* (London, UK: G.T. Foulis and CO Ltd, 1966).
6. Winnicker, C., Gaskill, B. N., Pritchett-Corning, K. & Garner, J. P. *A Guide to the Behavior & Enrichment of Laboratory Rodents* (Wilmington, MA: Charles River, 2012).
7. Stanford. *Mouse Ethogram* <www.mousebehavior.org>
8. Jennings, M. et al. Refining rodent husbandry: The mouse: Report of the Rodent Refinement Working Party. *Laboratory Animals* **32**, 233–259 (1998).
9. Gray, J. A. *The Psychology of Fear and Stress.* (2nd edition) (Cambridge: Cambridge University Press, 1991).
10. Weiss, J. Psychological factors in stress and disease. *Scientific American* **226**, 104–113 (1972).
11. Cannon, B. & Nedergaard, J. Nonshivering thermogenesis and its adequate measurement in metabolic studies. *Journal of Experimental Biology* **214**, 242–253 (2011).
12. Gaskill, B. N. & Pritchett-Corning, K. R. Nest building as an indicator of illness in laboratory mice. *Applied Animal Behaviour Science* **180**, 140–146 (2016). doi: 10.1016/j.applanim.2016.04.008
13. Jirkof, P. Burrowing and nest building behavior as indicators of well-being in mice. *Journal of Neuroscience Methods* 234, 139–146 (2014). doi: 10.1016/j.jneumeth.2014.02.001
14. Oliver, V. L., Thurston, S. E. & Lofgren, J. L. Using cageside measures to evaluate analgesic efficacy in mice (*Mus musculus*) after Surgery. *Journal of the American Association for Laboratory Animal Science* **57**, 186–201 (2018).
15. Sealander, J. A. The relationship of nest protection and huddling to survival of *Peromyscus* at low temperature. *Ecology* **33**, 63–71 (1952).
16. Lynch, G. R., Lynch, C. B., Dube, M. & Allen, C. Early cold exposure: Effects on behavioral and physiological thermoregulation in the house mouse, *Mus musculus. Physiological Zoology* **49**, 191–199 (1976).
17. Lynch, C. B. & Possidente, B. P. Relationships of maternal nesting to thermoregulatory nesting in house mice (*Mus musculus*) at warm and cold temperatures. *Animal Behaviour* **26**, 1136–1143 (1978).
18. Gaskill, B. N., Lucas, J. R., Pajor, E. A. & Garner, J. P. Working with what you've got: Changes in thermal preference and behavior in mice with or without nesting material. *Journal of Thermal Biology* **36**, 1193–1199 (2011). doi: 10.1016/j.jtherbio.2011.02.004
19. Gaskill, B. N. et al. Heat or insulation: Behavioral titration of mouse preference for warmth or access to a nest. *PLoS ONE* **7**, e32799 (2012).
20. Gaskill, B. N. et al. Impact of nesting material on mouse body temperature and physiology. *Physiology & Behavior* **110**, 87–95 (2013). doi: 10.1016/j.physbeh.2012.12.018
21. Gaskill, B. N. et al. Energy reallocation to breeding performance through improved nest building in laboratory mice. *PLoS ONE* **8**, e74153 (2013).
22. Broida, J. & Svare, B. Strain typical patterns of pregnancy induced nest-building in mice - Maternal and experiential influences. *Physiology & Behavior* **29**, 153–157 (1982).
23. Aubert, A., Goodall, G., Dantzer, R. & Gheusi, G. Differential effects of lipopolysaccharide on pup retrieving and nest building in lactating mice. *Brain Behavior and Immunity* **11**, 107–118 (1997). doi: 10.1006/brbi.1997.0485

24. Gaskill, B. N., Karas, A. Z., Garner, J. P. & Pritchett-Corning, K. R. Nest building as an indicator of health and welfare. *Journal of Visualized Experiments* **82**, e51012 (2013).

25. Deacon, R. M. J. Burrowing in rodents: A sensitive method for detecting behavioral dysfunction. *Nature Protocols* **1** 118–121 (2006). doi: 10.1038/nprot.2006.19

26. Deacon, R. et al. Age-dependent and-independent behavioral deficits in Tg2576 mice. *Behavioural Brain Research* **189**, 126–138 (2008).

27. Deacon, R. M., Raley, J. M., Perry, V. H. & Rawlins, J. N. P. Burrowing into prion disease. *Neuroreport* **12**, 2053–2057 (2001).

28. Guenther, K., Deacon, R. M., Perry, V. H. & Rawlins, J. N. P. Early behavioural changes in scrapie-affected mice and the influence of dapsone. *European Journal of Neuroscience* **14**, 401–409 (2001).

29. Deacon, R. M. J., Croucher, A. & Rawlins, J. N. P. Hippocampal cytotoxic lesion effects on species-typical behaviours in mice. *Behavioural Brain Research* **132**, 203–213 (2002). doi: 10.1016/s0166-4328(01)00401-6

30. Contet, C., Rawlins, J. N. P. & Deacon, R. M. A comparison of 129S2/SvHsd and C57BL/6JOlaHsd mice on a test battery assessing sensorimotor, affective and cognitive behaviours: Implications for the study of genetically modified mice. *Behavioural brain research* **124**, 33–46 (2001).

31. Jirkof, P. et al. Burrowing behavior as an indicator of post-laparotomy pain in mice. *Frontiers in Behavioral Neuroscience* **4**, 165 (2010). doi: 10.3389/fnbeh.2010.00165

32. Jirkof, P. et al. Burrowing is a sensitive behavioural assay for monitoring general wellbeing during dextran sulfate sodium colitis in laboratory mice. *Laboratory Animals* **47**, 274–283 (2013). doi: 10.1177/0023677213493409

33. Würbel, H. The motivational basis of caged rodent's stereotypies. In Rushen, J. & Mason, G. (Eds.), *Stereotypic Animal Behaviour: Fundamentals and Applications to Welfare* (2nd edition), Ch. 4 (Wallingdford, UK: CABI, 2006), 86–120.

34. Nevison, C. M., Hurst, J. L. & Barnard, C. J. Why do male ICR(CD-1) mice perform bar-related (stereotypic) behaviour? *Behavioural Processes* **47**, 95–111 (1999).

35. Garner, J. P. et al. Reverse-translational biomarker validation of Abnormal Repetitive Behaviors in mice: An illustration of the 4P's modeling approach. *Behavioural Brain Research* **219**, 189–196 (2011).

36. Cooper, J. J., Ödberg, F. & Nicol, C. J. Limitations on the effectiveness of environmental improvement in reducing stereotypic behaviour in bank voles (*Clethrionomys glareolus*). *Applied Animal Behaviour Science* **48**, 237–248 (1996).

37. Garner, J. P., Dufour, B., Gregg, L. E., Weisker, S. M. & Mench, J. A. Social and husbandry factors affecting the prevalence and severity of barbering ('whisker trimming') by laboratory mice. *Applied Animal Behaviour Science* **89**, 263–282 (2004). doi: 10.1016/j.applanim.2004.07.004

38. Vieira, G. d. L. T., Lossie, A. C., Lay, D. C., Jr., Radcliffe, J. S. & Garner, J. P. Preventing, treating, and predicting barbering: A fundamental role for biomarkers of oxidative stress in a mouse model of Trichotillomania. *PLoS ONE* **12**, e0175222 (2017). doi: 10.1371/journal.pone.0175222

39. Marx, J., Brice, A. K., Boston, R. C. & Smith, A. L. Incidence rates of spontaneous disease in laboratory mice used at a large biomedical research institution. *Journal of the American Association for Laboratory Animal Science* **52**, 782–791 (2013).

40. Dufour, B. D. et al. Nutritional up-regulation of serotonin paradoxically induces compulsive behavior. *Nutritional Neuroscience* **13**, 256–264 (2010). doi: 10.1179/147683010x12611460764688

41. Adams, S. C., Garner, J. P., Felt, S. A., Geronimo, J. T. & Chu, D. K. A "pedi" cures all: Toenail trimming and the treatment of ulcerative dermatitis in mice. *PLoS ONE* **11**, e0144871 (2016).

42. Howerton, C. L., Garner, J. P. & Mench, J. A. Effects of running wheel-igloo enrichment on aggression, hierarchy linearity, and stereotypy in group-housed male CD-1 (ICR) mice. *Applied Animal Behaviour Science* **115**, 90–103 (2008).

43. Barnard, C. J., Behnke, J. M. & Sewell, J. Environmental enrichment, immunocompetence, and resistance to *Babesia microti* in male mice. *Physiology & Behavior* **60**, 1223–1231 (1996).

44. Weber, E. M., Algers, B., Hultgren, J. & Olsson, A. Pup mortality in laboratory mice – infanticide or not? *Acta Veterinaria Scandinavica* **55**, 1–8 (2013).

45. Weber, E. M., Hultgren, J., Algers, B. & Olsson, I. A. S. Do laboratory mouse females that lose their litters behave differently around parturition? *PLoS ONE* **11**, e0161238 (2016). doi: 10.1371/journal.pone.0161238

46. Weber, E., Algers, B., Würbel, H., Hultgren, J. & Olsson, I. Influence of strain and parity on the risk of litter loss in laboratory mice. *Reproduction in Domestic Animals* **48**, 292–296 (2013).

47. Olsson, I. A. S. et al. Understanding behaviour: The relevance of ethological approaches in laboratory animal science. *Applied Animal Behaviour Science* **81**, 245–264 (2003).

48. Gerlai, R. & Clayton, N. S. Analysing hippocampal function in transgenic mice: An ethological perspective. *Trends in Neurosciences* **22**, 47–51 (1999).
49. Krakenberg, V. et al. Technology or ecology? New tools to assess cognitive judgement bias in mice. *Behavioural Brain Research* **362**, 279–287 (2019).
50. Crabbe, J. C., Wahlsten, D. & Dudek, B. C. Genetics of mouse behavior: Interactions with laboratory environment. *Science* **284**, 1670–1672 (1999).
51. Chesler, E. J., Wilson, S. G., Lariviere, W. R., Rodriguez-Zas, S. L. & Mogil, J. S. Identification and ranking of genetic and laboratory environment factors influencing a behavioral trait, thermal nociception, via computational analysis of a large data archive. *Neuroscience & Biobehavioral Reviews* **26**, 907–923 (2002).
52. Gouveia, K. & Hurst, J. L. Optimising reliability of mouse performance in behavioural testing: The major role of non-aversive handling. *Scientific Reports* **7**, 44999 (2017).
53. Ryan, B. C., Young, N. B., Moy, S. S. & Crawley, J. N. Olfactory cues are sufficient to elicit social approach behaviors but not social transmission of food preference in C57BL/6J mice. *Behavioural Brain Research* **193**, 235–242 (2008).
54. Roberts, S. A. et al. Darcin: A male pheromone that stimulates female memory and sexual attraction to an individual male's odour. *BMC Biology* **8**, 75 (2010).
55. Garner, J. P., Thogerson, C. M., Wurbel, H., Murray, J. D. & Mench, J. A. Animal neuropsychology: Validation of the intra-dimensional extra-dimensional set shifting task for mice. *Behavioural Brain Research* **173**, 53–61 (2006).
56. Bodden, C. et al. Heterogenising study samples across testing time improves reproducibility of behavioural data. *Scientific Reports* **9**, 8247 (2019).
57. Hossain, S. M., Wong, B. K. Y. & Simpson, E. M. The dark phase improves genetic discrimination for some high throughput mouse behavioral phenotyping. *Genes Brain and Behavior* **3**, 167–177 (2004). doi: 10.1111/j.1601-183x.2004.00069.x
58. Roedel, A., Storch, C., Holsboer, F. & Ohl, F. Effects of light or dark phase testing on behavioural and cognitive performance in DBA mice. *Laboratory Animals* **40**, 371–381 (2006).
59. Panksepp, J. Neuroevolutionary sources of laughter and social joy: Modeling primal human laughter in laboratory rats. *Behavioural Brain Research* **182**, 231–244 (2007).
60. Dolensek, N., Gehrlach, D. A., Klein, A. S. & Gogolla, N. Facial expressions of emotion states and their neuronal correlates in mice. *Science* **368**, 89–94 (2020).
61. Panksepp, J. The basic emotional circuits of mammalian brains: Do animals have affective lives? *Neuroscience & Biobehavioral Reviews* **35**, 1791–1804 (2011).
62. Panksepp, J. & Panksepp, J. B. Toward a cross-species understanding of empathy. *Trends in Neurosciences* **36**, 489–496 (2013).
63. Tzschentke, T. M. Measuring reward with the conditioned place preference (CPP) paradigm: Update of the last decade. *Addiction Biology* **12**, 227–462 (2007). doi: 10.1111/j.1369-1600.2007.00070.x
64. Schechter, M. D. & Calcagnetti, D. J. Continued trends in the conditioned place preference literature from 1992 to 1996, inclusive, with a cross-indexed bibliography. *Neuroscience & Biobehavioral Reviews* **22**, 827–846 (1998).
65. Cannon, W. B. *Bodily Changes in Pain, Hunger, Fear and Rage: An Account of Recent Researches into the Function of Emotional Excitement* (New York: D Appleton & Company, 1929).
66. Panksepp, J. Aggression elicited by electrical stimulation of the hypothalamus in albino rats. *Physiology & Behavior* **6**, 321–329 (1971).
67. Siegel, A. *Neurobiology of Aggression and Rage* (Boca Raton, FL: CRC Press, 2004).
68. Weber, E. M., Dallaire, J. A., Gaskill, B. N., Pritchett-Corning, K. R. & Garner, J. P. Aggression in group-housed laboratory mice: Why can't we solve the problem? *Lab Animal* **46**, 157 (2017).
69. Pfaus, J. G., Kippin, T. E. & Coria-Avila, G. What can animal models tell us about human sexual response? *Annual Review of Sex Research* **14**, 1–63 (2003).
70. Pfaff, D. W. *Drive: Neurobiological and Molecular Mechanisms of Sexual Motivation* (Cambridge, MA: MIT Press, 1999).
71. Panksepp, J. *Affective Neuroscience: The Foundations of Human and Animal Emotions* (Oxford, UK: Oxford University Press, 2004).
72. Wöhr, M. & Schwarting, R. K. W. Affective communication in rodents: Ultrasonic vocalizations as a tool for research on emotion and motivation. *Cell and Tissue Research* **354**, 81–97 (2013). doi: 10.1007/s00441-013-1607-9
73. Held, S. D. E. & Špinka, M. Animal play and animal welfare. *Animal Behaviour* **81**, 891–899 (2011).
74. Loy, J. Behavioral responses of free-ranging rhesus monkeys to food shortage. *American Journal of Physical Anthropology* **33**, 263–271 (1970).

75. Pellis, S. M. & Pasztor, T. J. The developmental onset of a rudimentary form of play fighting in C57 mice. *Developmental Psychobiology* **34**, 175–182 (1999). doi: 10.1002/(sici)1098-2302(199904)34:3 <175::aid-dev2>3.0.co;2-#

76. Pellis, S. & Pellis, V. *The Playful Brain: Venturing to the Limits of Neuroscience* (London, UK: Oneworld Publications, 2013).

77. Gaskill, B. N. & Pritchett-Corning, K. R. The Effect of cage space on behavior and reproduction in Crl:CD1(Icr) and C57BL/6NCrl laboratory mice. *PLoS ONE* **10**, e0127875 (2015). doi: 10.1371/journal .pone.0127875

78. Richter, S. H., Kästner, N., Kriwet, M., Kaiser, S. & Sachser, N. Play matters: The surprising relation-ship between juvenile playfulness and anxiety in later life. *Animal Behaviour* **114**, 261–271 (2016). doi: 10.1016/j.anbehav.2016.02.003

79. Würbel, H. & Garner, J. P. Refinement of rodent research through environmental enrichment and sys-tematic randomizations. NC3Rs 9, 1–9 (2007).

80. Weiss, J. M. Somatic effects of predictable and unpredictable shock. *Psychosomatic Medicine* **32**, 397–408 (1970).

81. Cao, L. et al. Environmental and genetic activation of a brain-adipocyte BDNF/leptin axis causes cancer remission and inhibition. *Cell* **142**, 52–64 (2010). doi: 10.1016/j.cell.2010.05.029

82. Cao, L. et al. White to brown fat phenotypic switch induced by genetic and environmental activation of a hypothalamic-adipocyte axis. *Cell Metabolism* **14**, 324–338 (2011). doi: 10.1016/j.cmet.2011.06.020

83. Cao, L. et al. Molecular therapy of obesity and diabetes by a physiological autoregulatory approach. *Nature Medicine* **15**, 447–454 (2009). doi: 10.1038/nm.1933

84. Reinhardt, V. (ed.), *Committed to Animal Welfare*, Discussions by the Laboratory Animal Refinement & Enrichment Forum. Vol. IV (Washington, DC: Animal Welfare Institute, 2016).

85. Reinhardt, V. (ed.), *Caring Hands*, Discussions by the Laboratory Animal Refinement & Enrichment Forum. Vol. II (Washington, DC: Animal Welfare Institute, 2010).

86. National Research Council. *Guide for the Care and Use of Laboratory Animals* (8th edition) (Washington, DC: National Academies Press, 2010).

87. Directive, C. Directive 2010/75/EU of the European parliament and of the council. *The Official Journal of the European Union* L 334, 17–119 (2010).

88. Young, R. J. *Environmental Enrichment for Captive Animals* (Hoboken, NJ: Blackwell Science Ltd, 2003).

89. Humphrey, N. K. The social function of intellect, in Bateson, P. P. G. & Hinde, R. A. (Eds.), *Growing Points in Ethology* (Cambridge: Cambridge University Press, 1976), pp. 303–317.

90. Van Loo, P. L. P. et al. Impact of 'living apart together' on postoperative recovery of mice compared with social and individual housing. *Laboratory Animals* **41**, 441–455 (2007). doi: 10.1258/002367707782314328

91. D'Amato, F. R. Neurobiological and behavioral aspects of recognition in female mice. *Physiology & Behavior* **62**, 1311–1317 (1997).

92. Pham, T. M. et al. Housing environment influences the need for pain relief during post-operative recov-ery in mice. *Physiology & Behavior* **99**, 663–668 (2010). doi: 10.1016/j.physbeh.2010.01.038

93. Sherwin, C. M. The influences of standard laboratory cages on rodents and the validity of research data. *Animal Welfare* **13**, S9–S15 (2004).

94. Van Loo, P. L. P., Kruitwagen, C. L. J. J., Van Zutphen, L. F. M., Koolhaas, J. M. & Baumans, V. Modulation of aggression in male mice: Influence of cage cleaning regime and scent marks. *Animal Welfare* **9**, 281–295 (2000).

95. Morgan, K. N. & Tromborg, C. T. Sources of stress in captivity. *Applied Animal Behaviour Science* **102**, 262–302 (2007). doi: 10.1016/j.applanim.2006.05.032

96. Sherwin, C. M. Preferences of individually housed TO strain laboratory mice for loose substrate or tubes for sleeping. *Laboratory Animals* **30**, 245–251 (1996).

97. Mertens, S. et al. Effect of three different forms of handling on the variation of aggression-associated parameters in individually and group-housed male C57BL/6NCrl mice. *PLoS ONE* **14**, e0215367 (2019). doi: 10.1371/journal.pone.0215367

98. Aubert, A. Sickness and behaviour in animals: A motivational perspective. *Neuroscience and Biobehavioral Reviews* **23**, 1029–1036 (1999). doi: 10.1016/s0149-7634(99)00034-2

99. Jirkof, P. et al. Assessment of postsurgical distress and pain in laboratory mice by nest complexity scor-ing. *Laboratory Animals (London)* **47**, 153–161, doi:10.1177/0023677213475603 (2013).

100. Dewsbury, D. Studies of rodent-human interactions in animal psychology. In Davis, H. & Balfour, D. (Eds.), *The Inevitable Bond: Examining Scientist-Animal Interactions* (Cambridge: Cambridge University Press, 1992), pp, 27–43.

101. Flecknell, P. Small mammals. In Anderson, R.S. & Edney, A.T.B. (Eds.), *Practical Animal Handling* (Oxford, UK: Pergamon Press, 1991), pp. 177–188.
102. Hurst, J. L. & West, R. S. Taming anxiety in laboratory mice. *Nature Methods* **7**, 825–826 (2010).
103. Gouveia, K. & Hurst, J. L. Reducing mouse anxiety during handling: Effect of experience with handling tunnels. *PLoS ONE* **8**, e66401 (2013).
104. Hemsworth, P. H. Human–animal interactions in livestock production. *Applied Animal Behaviour Science* **81**, 185–198 (2003). doi: 10.1016/S0168-1591(02)00280-0
105. Chang, F. T. & Hart, L. A. Human-animal bonds in the laboratory: How animal behavior affects the perspectives of caregivers. *Ilar Journal* **43**, 10–18 (2002).
106. Balcombe, J. P. Laboratory environments and rodents' behavioural needs: A review. *Laboratory Animals* **40**, 217–235 (2006).
107. Shyan-Norwalt, M. R. The human-animal bond with laboratory animals. *Lab Animal* **38**, 132 (2009).
108. Wolfensohn, S. & Lloyd, M. *Handbook of Laboratory Animal Management and Welfare* (Hoboken, NY: John Wiley & Sons, 2008).
109. Anderson, R. S. & Edney, A. T. B. *Practical Animal Handling* (Oxford, UK: Pergamon Press, 1991).
110. Buerge, T. & Weiss, T. Handing and restraint. In Hendrich, H. J. & Bullock, G (Eds.), *The Laboratory Mouse. Handbook of Experimental Animals* Ch. 31 (Oxford: Academic Press Elsevier, 2004), pp. 517–526.
111. National Center for the Replacement Refinement & Reduction of Animals in Research. *Mouse Handling Research Papers*. Available at: https://www.nc3rs.org.uk/mouse-handling-research-papers
112. Clarkson, J. M., Dwyer, D. M., Flecknell, P. A., Leach, M. C. & Rowe, C. Handling method alters the hedonic value of reward in laboratory mice. *Scientific Reports* **8**, 2448 (2018).
113. Ono, M. et al. Does the routine handling affect the phenotype of disease model mice? *Japanese Journal of Veterinary Research* **64**, 265–271 (2016).
114. Ghosal, S. et al. Mouse handling limits the impact of stress on metabolic endpoints. *Physiology & Behavior* **150**, 31–37 (2015).
115. Henderson, L. J., Smulders, T. V. & Roughan, J. V. Identifying obstacles preventing the uptake of tunnel handling methods for laboratory mice: An international thematic survey. *PLoS ONE* **15**, e0231454 (2020). doi: 10.1371/journal.pone.0231454
116. Littin, K. et al. Towards humane end points: Behavioural changes precede clinical signs of disease in a Huntington's disease model. *Proceedings of the Royal Society Biological Sciences Series B* **275**, 1865–1874 (2008).
117. Gouveia, K. & Hurst, J. L. Improving the practicality of using non-aversive handling methods to reduce background stress and anxiety in laboratory mice. *Scientific Reports* **9**, 1–19 (2019).
118. Richter, S. H., Garner, J. P. & Wurbel, H. Environmental standardization: Cure or cause of poor reproducibility in animal experiments? *Nature Methods* **6**, 257–261 (2009).
119. André, V. et al. Laboratory mouse housing conditions can be improved using common environmental enrichment without compromising data. *PLoS Biology* **16**, e2005019 (2018).
120. Skinner, M., Ceuppens, P., White, P. & Prior, H. Social-housing and use of double-decker cages in rat telemetry studies. *Journal of Pharmacological and Toxicological Methods* **96**, 87–94 (2019).

9 The Rat

I. Joanna Makowska

CONTENTS

As a domesticated animal which may be readily gentled, it is responsive to care and attention and thrives under conditions of comfort and contentment. It soon becomes accustomed to, or even develops an interest in, activities about it, evinces an affection for its caretakers, and shows its pleasure in playful pranks not unlike other domestic animals.

– Greenman and Duhring (1923), *Breeding and Care of*
the Albino Rat for Research Purposes, **p. 12.**

INTRODUCTION

Norway rats are the first animals to have been domesticated primarily for scientific purposes.[1] They were brought into laboratories sometime in the 1840s,[2] and were first described as research subjects in a scientific article published in 1856.[3] In those days, rats were kept in "homemade" cages that were improvised by individual researchers.[4,5] These cages were similar in size and shape to those still used today; indeed, the design of today's standard laboratory cages has changed little since the early 1900s[6] and is based more on tradition than scientific evidence.[7–9]

Conventional laboratory rat cages are very different from the environments the animals have evolved in. Even though laboratory rats have been domesticated, the process of domestication does not eliminate any species-specific behaviors, although it may alter the quality or thresholds needed to initiate them.[10,11] Standard laboratory cages, due to their small size and simplicity, thwart rats' ability to express a range of natural behaviors they would perform regularly in less restrictive captive environments.[12]

In addition to the limitations imposed by the physical environment, rats in laboratories must also contend with artificial social environments – notably, coming into direct contact with humans. Domesticated rats, while more tolerant of humans compared to wild rats,[13] are still inherently fearful of humans and human touch.[14,15] With the right approach, however, rats can overcome their fear of humans and develop rewarding relationships with their caretakers.

This chapter draws on existing scientific literature and a few personal anecdotes to illustrate key features of a rat-centric laboratory environment. When attempting to design species-appropriate captive environments, we must be careful to consider who the animals are, rather than who we assume them to be. Observations of free-ranging animals in their natural habitats, where they have ample resources and opportunities to make choices, can offer important insights into what is important to them, and under which set of circumstances. Indeed, observations of natural behaviors often form the basis for testable hypotheses within laboratory settings.[16] However, when interpreting the results of preference or motivation tests, we must be cognizant that *what* we ask the animals, and *how* we ask it, invariably determine the types of answers we receive.[17]

FREE-RANGING RATS: WHO ARE THEY?

The physical world of free-ranging *Rattus norvegicus* consists at its core of a home burrow, one or more food sources, and a series of narrow pathways connecting the two. Burrows are made up of one or more spherical nest chambers lined with plant materials, a few food caches, and many connecting tunnels.[18–21] A typical burrow system has several open entrances and a few "bolt holes" that are loosely covered with vegetation. Although Norway rats live in colonies of several hundred individuals, they segregate into smaller social groups with which they share a home burrow. These groups usually approximate the individuals' litter size, which is in the range of 5–12 individuals.[19]

Rats emerge from their burrows soon after waking up to forage, socialize, patrol their territory, or explore new grounds.[22,23] Their home range, which is largely determined by the distance between the burrow and the food sources,[24–26] is approximately 180–360 square yards in urban and farm settings[24–26] and 7500–69,100 square yards in fields and grasslands.[25,27] Rats usually travel close to cover (e.g., along hedgerows and fences or under vegetation) along well-worn pathways,[18,22,23] but will sometimes venture out into the open for longer periods; when they do this, they are likely to excavate a short burrow they can retreat to if disturbed.[22]

Rats are active outside of their burrows for 5–11 hours per day, depending on the season and study site.[19,22,23,26] They adjust the time at which they are outside of their burrows to avoid peaks in human activity. For example, rats living around a busy market in Malaysia retreated back into their burrows just as the market was set to open.[26] On a university campus in California, rats who were mostly nocturnal exhibited a gradual shift towards diurnality in the weeks when fewer students were around, and abruptly reversed back to nocturnality when the students returned.[23]

HOUSING FOR RATS IN LABORATORIES

Laboratory animals spend the majority of their life within a cage. Therefore, the features and characteristics of this cage largely define the breadth and scope of their life experiences and are thus important factors affecting their welfare.

A good housing environment would provide rats with access to resources and opportunities that are important to them.[28] Moreover, a good housing environment would also allow animals to exercise some control over their environment and what happens to them. Indeed, the ability to be agentic is recognized to be fundamental to a good life. It is suggested that, unlike exerting behavior simply in direct reaction to external stimuli (*passive/reactive agency*), engaging in active behavior to achieve outcomes (*action-driven agency*; e.g., building a nest or a burrow) and active skill building and information acquisition for later use (*competence-building agency*; e.g., play or exploration) are associated with positive affective states.[29] Good welfare is also related to the ability to pursue one's own interests, engage in intentional actions, and live up to one's full potential through the utilization of one's innate abilities.[30–32] Housing designed with the animal as a central consideration would thus not only allow animals to fulfill important, species-specific behavioral functions, but also provide opportunities to make choices with regards to how, when, and where to do so.

A Historical Perspective

In the early 1900s, as the use of rats as research models was beginning to gain momentum, scientists at the Wistar Institute of Anatomy and Biology in Philadelphia took a different approach than most of their contemporaries: instead of improvising cages for rats, they wanted to gain an "intimate acquaintance with the habits of this little animal" to uncover "the means of making it contented and happy." Thus, in 1906, institute director Milton J. Greenman and his assistant F. Louise Duhring began to study the conditions necessary for maintaining biologically and psychologically healthy rats with the goal of developing appropriate housing and husbandry.[33] Their findings are described in their book *Breeding and Care of the Albino Rat for Research Purposes*, which was published in 1923.[4]

Some of Greenman and Duhring's research findings were that "confining a rat to the limited quarters of a cage necessarily restricts its activities, modifies its mental processes, and influences its growth and development," and that "fear and lack of exercise are factors which react unfavorably upon the growing rat." To compensate, "as far as is possible, for the disadvantages of cage life," they designed two types of wooden cages that were to be used in combination: a *dormer* cage (from the French dormir, to sleep) and an *exercising* cage.[4]

The dormer cage measured 17" W × 35" L × 12.5" H and housed up to ten rats. The cage was divided into two compartments by means of a partition with a circular opening; the purpose of the division was to segregate space so it could be used for different activities, and to offer rats the opportunity to cross to the adjacent compartment if they became frightened. It was noted that "this simple shifting of location appears to satisfy the animal that it has protected itself." The cage was furnished with bedding and nesting material (wood wool/excelsior was preferred) in which rats formed burrows and nests. The exercising cage was identical, except that it connected to a running wheel that measured 21" in diameter. Females with litters were sometimes seen taking individual pups in their mouth and running in the wheel with them one by one, until the whole litter had been with her in the wheel. Greenman and Duhring deemed the exercising cage to be "an essential part of the colony equipment." [4]

Current Knowledge

Greenman and Duhring's early studies offered important insights into what factors in the animals' physical environment are important to rats. More recent research, using more sophisticated methods of assessing animal welfare, has also identified some of these factors as being of high importance to the animals' well-being, namely: access to a shelter, preferably a burrow; exercising and mobility; and environmental complexity that allows spatial and social segregation. These factors are described in more detail below.

Shelters

Rats are burrowing animals with a strong motivation to hide. Providing rats with an adequate shelter is essential, as it allows them to control their exposure to perceived threats, bright lights, and ambient temperatures.[12,34,35] Rats not only prefer cages with a shelter,[35–38] but are also less anxious when provided with a shelter.[36] Rats prefer shelters that are dark, opaque, and made of Plexiglas over shelters that are clear or made of cardboard or tin.[39] Importantly, rats prefer shelters with one entrance rather than open-ended tunnels.[37,39–41]

Rats are also motivated to access nesting material, which they drag inside the shelter, manipulate, and rest on top of[37,39,41] (this is akin to lining their nesting chamber with various plant materials in the wild[18,19,21]). Rats prefer long and wide strips of paper (e.g., sheets of brown paper towel[41] or soft paper strips[37]) or straw[42] over paper or wood shavings.[37] They prefer to have both a nest box and nesting material, but prefer only a nest box versus only nesting material.[38]

Norway rats evolved to live in underground burrows,[18] and they excavate burrows even when man-made nest boxes are provided.[19] Laboratory rats have retained their motivation to excavate burrows (Figure 9.1).[12,43,44] Recently, Makowska and Weary[12] showed that Sprague-Dawley rats housed in a multi-level cage containing nest boxes and soil worked on burrow excavation approximately 30 times per day. Rats maintained this burrowing rate as they grew older, even though they decreased the rate of other active behaviors, such as climbing and exploration. This finding suggests that burrowing is particularly important to rats. These rats spent the majority of the light period inside their burrows, and emerged during the dark period to explore their cage, climb, socialize, excavate burrows, and rest in a hammock or nest box.

Building and retreating into a shelter may be one of the few opportunities captive rodents have to actively engage with their environment. Excavating a burrow and manipulating nesting material allow the animals to create something they will later use, which constitutes an example of Špinka's[29] *action-driven agency* and is thus likely associated with positive affective states. The idea that rats enjoy excavating burrows is supported by the finding that they continued to excavate new tunnels even when the old ones were still present; i.e., burrow excavation per se was important, irrespective of the end goal of having a burrow.[12]

Exercising and Mobility

There is evidence that rats are motivated to engage in physical activity for its own sake. Indeed, rats with access to a running wheel will use it extensively; in one study, rats with voluntary access to a running wheel used it to travel a distance equivalent to approximately 0.6 miles per day.[45] It has been suggested that wheel running by captive rodents is an unnatural, abnormal behavior or stereotypy.[46,47] However, wild rats and other small animals were found to use a wheel – seemingly on purpose – when it was encountered in the wild.[48] Laboratory rats are highly motivated to perform wheel running,[46] and it is an activity they seem to also enjoy: they display 50 kHz ultrasonic vocalizations (believed to indicate positive affective states[49]) both in anticipation of, and during the use of, a running wheel.[50]

Rats housed in spacious cages not only engage in a variety of species-specific, active behaviors, but also exhibit more indicators of positive welfare. For example, rats housed in the multi-level cages with burrowing substrate described above also frequently climbed (about 75 times per day

FIGURE 9.1   The burrows of Sprague-Dawley rats housed in large cages with soil. Left: four-week-old rat peeking out of a burrow, approximately 18 hours after being placed in a cage with soil for the first time. Right: rats living with burrowing soil excavated new tunnels on a daily basis. Photos by I. Joanna Makowska.

at three months old) and stood upright with their hind limbs and backs fully elongated (about 120 times per day at three months old).[12] These rats had better welfare compared to control standard-housed rats.[51] In a different study, groups of rats were housed in two multi-level ferret cages connected by a tube, and furnished with multiple tunnels, a large nest box, and wood chew blocks.[52] These rats expressed species-specific hopping gait, climbing, jumping, and extensive foraging; and they were also calmer, easier to handle, and did not startle easily. Finally, rats housed in a multi-level modified ferret rack engaged in spontaneous active behavior, such as jumping, running through the staircase, and exploratory behavior; these rats were also easier to handle by their caretakers.[53] In general, rats housed in larger cages that allow the expression of various active behaviors have better welfare, although this finding is often confounded with a simultaneous increase in group size and environmental complexity (see next subsection).

Providing rats with opportunities to locomote and exercise – ideally through the provision of an environment that allows various forms of activity like climbing, running, and exploration – is also recommended for their physical health. Indeed, standard-housed (sedentary) rodents have compromised metabolisms, which causes them to be overweight, insulin resistant, hypertensive, and with physiological profiles consistent with increased disease susceptibility.[54] Moreover, it has been suggested that low mobility, including the inability to stand and stretch upright, may lead standard-housed rats to experience physical discomfort: standard-housed rats performed eight times more lateral stretching than age-matched rats living in large cages, presumably because frequent stretching helped to alleviate joint stiffness and positional stress resulting from low mobility, including the inability to stretch upright.[12]

Complexity

Rats prefer structurally complex environments,[39,55] and also display more indicators of positive welfare when housed in these more complex environments.[51,56–59] Several literature reviews have shown that rats housed in "enriched" environments have lower anxiety and depressive-like states,[60–62] stronger immune systems,[60] and lower susceptibility to neurodegenerative diseases[63,64] and cancers.[65] In these reviews, "enriched" usually denotes any environment that is not barren. Complex and/or more biologically relevant laboratory environments are also associated with lower incidence of abnormal behaviors.[47,66,67]

Complexity usually consists of adding structures or objects to the cage that the animals can use or interact with; for example, shelters, tunnels, multiple levels, or climbing structures. There is some evidence that rats value not only the presence of several objects within their cage, but also prefer these objects to be diverse. For example, rats housed in standard-sized cages with five different objects in their cage had more indicators of good welfare and fewer indicators of poor welfare than rats housed either with any one of those same five objects,[56] or with five copies of any one of those five objects.[58]

A simple means of increasing complexity within a standard-sized cage is to add vertical dividers. Although this practice appears to be more common with laboratory mice,[68–70] it may also be beneficial to rats. Anzaldo and colleagues[71] found that group-housed male rats preferred cages with two L-shaped vertical dividers. The dividers increased opportunities for thigmotaxis, allowing rats to huddle and sleep together while maintaining contact with the walls. A more effective way to increase complexity is to provide multiple usable objects; these objects not only facilitate thigmotaxis, but also help to alleviate boredom[72–74] by providing some opportunities for exploration and interaction with the environment.[55,75] Objects also help to break up the space into distinct areas, facilitating both social and spatial segregation. Social segregation (separating from cage mates) may help to de-escalate aggressive interactions,[76] while spatial segregation (dividing the cage into functionally different areas) enables rats to better establish separate sites for resting and elimination.[18,55,77–79]

RECOMMENDATIONS FOR IMPLEMENTATION

An environment designed around rats' welfare would be spacious enough to allow them to run, hop, play, and explore. The environment would also contain nesting material (preferably long paper strips) and a blind-ended (versus open-ended) dark nest box large enough to accommodate all rats at once. The cage would be furnished with a variety of usable objects, such as chewing blocks, tunnels, ladders, running wheels, and burrowing substrate. These objects would not only facilitate thigmotaxis and social and spatial segregation, but also provide opportunities for different forms of exploration and interaction with the environment. With regards to burrowing substrate, soil (which can be autoclaved and sterilized) may be ideal from the rats' point of view,[43] but it may be difficult to maintain in a laboratory. To help with upkeep, soil can be contained within a smaller box with a partial lid inside the animals' cage. Wood wool (excelsior) can also be used as a substitute for soil (Figure 9.2).

While keeping rats in this way should be the goal, the transition to this type of housing may be slow due to practical constraints. In this case, one solution is to implement playpens: rats live in standard cages but are given regular access to an area that provides them with opportunities to engage with a complex environment. For example, Shenton[80] modified a spare rabbit cage for use with rats at her facility. The cage was furnished with items that she found within the research facility – empty glove boxes, tunnels, paper towel rolls, bins filled with crinkle paper or shallow water – and transferred four to six groups of pair-housed male rats into this "playpen" three times per week for up to several hours. Similarly, our department has been using playpens for rats with great success (Figure 9.3). The rats show positive anticipation on their way to the playpens and use their time in the playpens to run around, explore the different levels and objects, and engage in rough-and-tumble play. We, and others who use playpens, have noticed that these animals become easier to work with, calmer, and more interactive when handled.[80,81]

Rat playpens can be made by re-purposing cages for larger species or custom-building something to fit a specific facility (e.g., floor pen). Only one or two playpens are needed, because animals use them temporarily. Playpens can be cleaned between groups, but when such precautions are not necessary, we have found that rats enjoy investigating the smells of the other individuals who had used the playpen before them (unpublished data). While transferring rats to and from the playpen, caretakers can take the opportunity to interact with the animals in a playful manner – the importance of this is described in the following section.

FIGURE 9.2 Burrowing substrate can be contained within a smaller bin inside the animals' cage. Wood wool (excelsior) does not allow rats to excavate elaborate tunnels the way that soil does, but rats can nonetheless tunnel and nest within it. Photos by Anna Ratuski.

FIGURE 9.3 Playpen used by PhD candidate Lucía Améndola at the University of British Columbia. These rats, who were used in a study investigating personality differences in response to carbon dioxide, were given access to a playpen several times per week as environmental enrichment. Photo by I. Joanna Makowska.

RELATIONSHIPS WITH RATS IN LABORATORIES

Like many species, rats have an innate fear of humans. In laboratories, it is common for researchers and staff to interact with rats only during handling and restraint required by routine husbandry and experimental procedures. However, laboratory rats whose interactions with humans are limited to routine husbandry procedures, such as weighing and cage changing, remain anxious about these interactions. For example, such rats were found to produce 22 kHz vocalizations indicative of anxiety after being touched by a human[14] and to show increases in corticosterone,[82,83] noradrenaline,[82] heart rate, and blood pressure[84–86] when gently lifted from their cages.

In a good living environment, rats would not experience repeated and/or ongoing anxiety resulting from benign, routine procedures. Fortunately, with the right approach, rats can not only overcome their apprehension about humans, but also develop a rewarding relationship with their caretakers. Indeed, rats can become so friendly that a veterinarian once confessed to me that when clients told him they wanted a very, very small dog, he would advise them to get a rat instead.

A Historical Perspective

The importance of developing positive relationships with laboratory rats was uncovered in the early 1900s. Greenman and Duhring,[4] whose early work on appropriate rat housing was described in the previous section, also discovered the importance of fostering a positive relationship with laboratory rats. The researchers found that "individual attention, shown by handling and petting, is essential for the best growth of albino rats and for securing uniform reactions when used as research animals." For example, they wrote that survival after the removal of the parathyroid gland was only 15% among unsocialized rats, but 75% among socialized rats; and that the response of unsocialized rats' intestinal muscle segment following sodium carbonate stimulation was irregular or in the opposite way compared to that of socialized rats.

Greenman and Duhring[4] found that rats quickly learned to distinguish "kindness and attention" from humans and emphasized the importance of selecting caretakers with these qualities. They wrote that a good caretaker was "the person who will talk to them while working, and occasionally open a cage door, smooth the rats, induce them to romp and play, or pick out a number at a time and place them on their shoulders ... ." With the right caretaker, they argued, rats would "jump and frolic, using the front feet to dab playfully at an attendant known to them and one who will play with them." The authors concluded that "taming the colony or maintaining it in a condition of fearless contentment is not only economically desirable, but scientifically essential."

CURRENT KNOWLEDGE

When socializing rats to humans, it may be beneficial to start early. Rats enter a "stress hyporesponsive period," characterized by low baseline glucocorticoid levels and a reduced response to stressors, between approximately postnatal day four and postnatal day 14 (reviewed by Raineki and colleagues[87]). Rats subjected to "neonatal handling" during this period are more playful[88] and show reduced stress responses and faster return to baseline levels following a variety of stressors, including restraint, later in life.[87] "Neonatal handling" is an experimental paradigm that consists of picking up pups and placing them individually in a small compartment for several minutes. The long-term anxiolytic effects may be due to changes in maternal care after the separation, the handling itself, or a combination of the two.[89,90]

Regardless, gentle interaction with juvenile or adolescent rats also reduces rats' apprehension towards humans. Several approaches may be employed. Maurer and colleagues[91] developed a gentling program for newly weaned rats. For two weeks immediately after weaning, twice a day for ten minutes, rats were gently touched on their entire bodies, including tails. During each session, rats were also briefly lifted twice, hand-fed treats, and talked to in a friendly and soothing manner. Compared to control rats who were only interacted with during routine husbandry procedures, gentled rats were less likely to squeak, freeze, or run away when approached or lifted by both familiar and unfamiliar persons. Control rats eventually habituated to humans, but not until the animals were six months old.

Similarly, Costa and colleagues[92] placed 60-day-old rats on their lap or on a table and gently stroked the animals' necks and backs for five minutes, five times per week for six weeks. These rats had lower basal norepinephrine levels and lower anxiety in the elevated plus-maze compared to control rats who were only handled during routine husbandry procedures.

Greenman and Duhring's[4] early studies emphasized the importance of playful interactions between rats and humans. Indeed, it has more recently been discovered that the most effective way to socialize some rats to humans may be to "tickle" them. This approach, also called "playful handling" or "heterospecific play," consists of using a human hand to simulate juvenile rat rough-and-tumble play. Specifically, handlers make light, brisk, and vigorous movements with their fingertips on the rat's neck and stomach.[93] A systematic review of rat tickling concluded that, compared to rats who experience a passive hand, light touching, or minimal touching, rats who are tickled are more likely to approach humans and are less likely to experience fear and anxiety when handled.[94] Tickling also reduces rats' anxiety towards other types of stressors: for example, tickling rats before an intraperitoneal injection reduces the negative experience of the injection.[95,96] Tickling may be more effective when first applied to juvenile rather than adult rats.[94]

In addition to reducing fear and anxiety towards humans and other stressors, tickling also appears to be something that rats enjoy. Indeed, many rats emit positive 50 kHz vocalizations in anticipation of and during tickling (reviewed by LaFollette and colleagues[94]), as well as when exposed to an odor associated with tickling.[97] Rats also spend more time in a location where they have been tickled, and learn to press a lever for a tickling reward.[98] Compared to control rats who are simply handled, tickled rats who vocalize in the 50 kHz range during tickling are also more optimistic[99] (this is believed to indicate they are in a more positive emotional state[100]). Recent work has shown that 15 seconds of tickling for three days may be sufficient to increase some indicators of positive welfare in rats.[101]

RECOMMENDATIONS FOR IMPLEMENTATION

Humans are a constant presence in the lives of laboratory rats. Rather than causing them anxiety, we can become an important source of enrichment for these animals. Brief handling of neonates, coupled with as few as three 15-second, short tickling sessions, may be an effective and efficient way to socialize rats and reduce their negative responses to a variety of stressors.

That notwithstanding, it is important to continue to engage in positive social interactions with rats beyond the few interactions sufficient to decrease their negative responses to stressors. Indeed, rats experience positive affective states with each playful interaction. Playful interactions are likely valuable because they are dynamic and transactional: the two individuals engage in spontaneous behavior that the other has to respond to.[102] Joyce Sato-Reinhold, who was a visiting researcher at the University of British Columbia at the time, would let her research rats into a playroom for about an hour each day. While staying in the playroom to watch over the animals, she would sometimes practice various yoga poses – and soon discovered that she (and her shifting body positions) were the greatest source of interest for the rats. The animals would climb on her and seemingly enjoy the challenge of staying balanced as she moved. She also dabbed playfully at the rats, who responded with 50 kHz vocalizations that could be heard using a bat detector. Recent work has also shown that laboratory rats who were taught to play hide-and-seek with humans were not only very good at this game – for example, when it was their turn to hide, they made fewer vocalizations and were more likely to choose opaque vs. transparent cover – but they were also eager to play and made "joy jumps" during the game.[103]

CONCLUSION

Rats are intelligent and playful animals who enjoy opportunities to engage with their physical and social environments. Rat-centric care and management would see animals in environments large and complex enough to allow them to perform a variety of natural active behaviors, including exploration, shelter building, and segregation of space. These rats would be socialized to humans from a young age and engaged with playfully by their caretakers.

Keeping rats in these conditions minimizes their experience of negative emotional states and gives them opportunities to experience positive emotional states. Indeed, rats are capable of experiencing a range of emotions, including fear, anxiety, depression, joy, and excitement,[104] as well as relief (e.g., sighing when an expected negative stimulus does not occur[105]), and empathy (e.g., helping another rat at a cost to themselves,[106-108] or sensitization to pain when witnessing a familiar rat in pain[109]). Interacting with rats in a playful manner and keeping them in conditions where they can act as rats also makes them more interesting to work with – which, in turn, further motivates us to better attend to their care and welfare.

REFERENCES

1. Richter, C. P. Domestication of the Norway rat and its implications for the problem of stress. *Association for Research in Nervous and Mental Disease* **29**, 19–47 (1949).

2. Richter, C. P. Rats, man, and the welfare state. *American Psychologist* **14**, 18–28 (1959).

3. Philipeaux, J. M. Note sur l'extirpation des capsules surrénales chez les rats albinos (Mus ratus). *Comptes Rendus l'Académie des Science* **43**, 904–906 (1856).

4. Greenman, M. J. & Duhring, F. L. *Breeding and Care of the Albino Rat for Research Purposes* (Philadelphia, PA: The Wistar Institute of Anatomy and Biology, 1923).

5. Hessler, J. R. The history of environmental improvements in laboratory animal science: Caging systems, equipment, and facility design. In McPherson, C. & Mattingly, S. (Eds.), *Fifty Years of Laboratory Animal Science* (Memphis, TN: American Association for Laboratory Animal Science, 1999), pp. 92–120.

6. Galef, B. G. & Durlach, P. Should large rats be housed in large cages? An empirical issue. *Canadian Psychology* **34**, 203–207 (1993).

7. Gaskill, B. N. & Pritchett-Corning, K. R. The effect of cage space on behavior and reproduction in Crl:CD(SD) and BN/Crl laboratory rats. *Journal of the American Association for Laboratory Animal* **54**, 1–10 (2015).

8. Scharmann, W. Improved housing of mice, rats and guinea-pigs: A contribution to the refinement of animal experiments. *ATLA* **19**, 108–114 (1991).

9. van de Weerd, H. A., Baumans, V., Koolhaas, J. M. & van Zutphen, L. F. M. Strain specific behavioural response to environmental enrichment in the mouse. *Journal of Experimental Animal Science* **36**, 117–127 (1994).

10. Price, E. O. Behavioral development in animals undergoing domestication. *Applied Animal Behaviour Science* **65**, 245–271 (1999).

11. Künzl, C., Kaiser, S., Meier, E. & Sachser, N. Is a wild mammal kept and reared in captivity still a wild animal? *Hormones and Behavior* **43**, 187–196 (2003).

12. Makowska, I. J. & Weary, D. M. The importance of burrowing, climbing and standing upright for laboratory rats. *Royal Society Open Science* **3**, 160136 (2016).

13. Blanchard, R. J., Flannelly, K. J. & Blanchard, D. C. Defensive behavior of laboratory and wild Rattus norvegicus. *Journal of Comparative Psychology* **100**, 101–107 (1986).

14. Brudzynski, S. M. & Ociepa, D. Ultrasonic vocalization of laboratory rats in response to handling and touch. *Physiology & Behavior* **52**, 655–660 (1992).

15. Gärtner, K. et al. Stress response of rats to handling and experimental procedures. *Laboratory Animals* **14**, 267–274 (1980).

16. Weeks, C. A. & Nicol, C. J. Behavioural needs, priorities and preferences of laying hens. *World's Poultry Science Journal* **62**, 296–308 (2006).

17. Franks, B. What do animals want? *Animal Welfare* **28**, 1–10 (2019).

18. Pisano, R. G. & Storer, T. I. Burrows and feeding of the Norway rat. *Journal of Mammalogy* **29**, 374–383 (1948).

19. Calhoun, J. B. *The Ecology and Sociology of the Norway Rat* (Public Health Service Publication No. 1008, 1963).

20. Nieder, L., Cagnin, M. & Parisi, V. Burrowing and feeding behaviour in the rat. *Animal Behaviour* **30**, 837–844 (1982).

21. Eibl-Eibesfeldt, I. The interactions of unlearned behaviour patterns and learning in mammals. In Fessard, A., Gerard, R. W., Konorski, J. & Delafresnaye, J. F. (Eds.), *Brain Mechanisms and Learning* (Springfield, IL: Blackwell Scientific Publications, 1961), pp. 53–73.

22. Taylor, K. D. Range of movement and activity of common rats (*Rattus norvegicus*) on agricultural land. *Journal of Applied Ecology* **15**, 663–677 (1978).

23. Recht, M. A. The fine structure of the home range and activity pattern of free-ranging telemetered urban Norway rats Rattus norvegicus (Berkenhout). *Bulletin of the Society of Vector Ecologists* **7**, 29–35 (1982).

24. Davis, D. E., Emlen, J. T. & Stokes, A. W. Studies on home range in the brown rat. *Journal of Mammalogy* **29**, 207–225 (1948).

25. Lambert, M. S., Quy, R. J., Smith, R. H. & Cowan, D. P. The effect of habitat management on home-range size and survival of rural Norway rat populations. *Journal of Applied Ecology* **45**, 1753–1761 (2008).

26. Oyedele, D. T., Sah, S. A. M., Kairuddin, L. & Wan Ibrahim W. M. M., Range measurement and a habitat suitability map for the Norway rat in a highly developed urban environment. *Tropical Life Sciences Research* **26**, 27–44 (2015).

27. Bramley, G. N. Home ranges and interactions of kiore (*Rattus exulans*) and Norway rats (*R. norvegicus*) on Kapiti Island, New Zealand. *New Zealand Journal of Ecology* **38**, 1–7 (2014).

28. Learmonth, M. J. Dilemmas for natural living concepts of zoo animal welfare. *Animals* **9**, 318 (2019).
29. Špinka, M. Animal agency, animal awareness and animal welfare. *Animal Welfare* **28**, 11–20 (2019).
30. Wemelsfelder, F. How animals communicate quality of life: The qualitative assessment of behaviour. *Animal Welfare* **16**, 25–31 (2007).
31. Purves, D. & Delon, N. Meaning in the lives of humans and other animals. *Philosophical Studies* **175**, 317–338 (2018).
32. Nussbaum, M. C. Beyond "Compassion and Humanity". In Sunstein, C. R. & Nussbaum, M. C. (Eds.), *Animal Rights: Current Debates and New Directions* (Oxford, UK: Oxford University Press, 2005), pp. 299–320.
33. Lindsey, J. R. Historical Foundations. In Baker, H. J., Lindsey, J. R. & Weisbroth, S. H. (Eds.), *The Laboratory Rat* (Cambridge, MA: Academic Press Inc., 1979), pp 1–36.
34. Van de Weerd, H. A., Van Loo, P. L. P., Van Zutphen, L. F. M., Koolhaas, J. M. & Baumans, V. Preferences for nest boxes as environmental enrichment for laboratory mice. *Animal Welfare* **7**, 11–25 (1998).
35. Patterson-Kane, E. G. Shelter enrichment for rats. *Contemporary Topics in Laboratory Animal Science* **42**, 46–48 (2003).
36. Townsend, P. Use of in-cage shelters by laboratory rats. *Animal Welfare* **6**, 95–103 (1997).
37. Manser, C. E., Broom, D. M., Overend, P. & Morris, T. H. Investigations into the preferences of laboratory rats for nest-boxes and nesting materials. *Laboratory Animals* **32**, 23–35 (1998).
38. Manser, C. E., Broom, D. M., Overend, P. & Morris, T. H. Operant studies to determine the strength of preference in laboratory rats for nest-boxes and nesting materials. *Laboratory Animals* **32**, 36–41 (1998).
39. Patterson-Kane, E. G., Harper, D. N. & Hunt, M. The cage preferences of laboratory rats. *Laboratory Animals* **35**, 74–79 (2001).
40. Chmiel, D. J. J. & Noonan, M. Preference of laboratory rats for potentially enriching stimulus objects. *Laboratory Animals* **30**, 97–101 (1996).
41. Bradshaw, A. L. & Poling, A. Choice by rats for enriched versus standard home cages: Plastic pipes, wood platforms, wood chips, and paper towels as enrichment items. *Journal of the Experimental Analysis of Behavior* **55**, 245–250 (1991).
42. Jegstrup, I. M., Vestergaard, R., Vach, W. & Ritskes-Hoitinga, M. Nest-building behaviour in male rats from three inbred strains: BN/HsdCpb, BDIX/OrlIco and LEW/Mol. *Animal Welfare* **14**, 149–156 (2005).
43. Boice, R. Burrows of wild and albino rats: Effects of domestication, outdoor raising, age, experience, and maternal state. *Journal of Comparative and Physiological Psychology* **91**, 649–61 (1977).
44. Price, E. O. Burrowing in wild and domestic Norway rats. *Journal of Mammalogy* **58**, 239–240 (1977).
45. Fonseca, I. A. T. et al. Exercising for food: Bringing the laboratory closer to nature. *The Journal of Experimental Biology* **217**, 3274–3281 (2014).
46. Sherwin, C. M. Voluntary wheel running: A review and novel interpretation. *Animal Behaviour* **56**, 11–27 (1998).
47. Mason, G., Clubb, R., Latham, N. & Vickery, S. Why and how should we use environmental enrichment to tackle stereotypic behaviour? *Applied Animal Behaviour Science* **102**, 163–188 (2007).
48. Meijer, J. H. & Robbers, Y. Wheel running in the wild. *Proceedings of the Royal Society B: Biological Sciences* **281**, 20140210 (2014).
49. Knutson, B., Burgdorf, J. & Panksepp, J. Ultrasonic vocalizations as indices of affective states in rats. *Psychological Bulletin* **128**, 961–977 (2002).
50. Heyse, N. C., Brenes, J. C. & Schwarting, R. K. W. Exercise reward induces appetitive 50-kHz calls in rats. *Physiology & Behavior* **147**, 131–140 (2015).
51. Makowska, I. J. & Weary, D. M. Differences in anticipatory behaviour between rats (*Rattus norvegicus*) housed in standard versus semi-naturalistic laboratory environments. *PLoS ONE* **11**, e0147595 (2016).
52. Clarke, D. & Ioannou, L. Introduction of gang caging for group housed rats. *Animal Technology and Welfare* **17**, 136–137 (2018).
53. Brenneis, C. et al. Automated tracking of motion and body weight for objective monitoring of rats in colony housing. *Journal of the American Association for Laboratory Animal* **56**, 18–31 (2017).
54. Martin, B., Ji, S., Maudsley, S. & Mattson, M. P. 'Control' laboratory rodents are metabolically morbid: Why it matters. *Proceedings of the National Academy of Sciences of the United States of America* **107**, 6127–6133 (2010).

55. Denny, M. S. The rat's long-term preference for complexity in its environment. *Animal Learning & Behavior* **3**, 245–249 (1975).

56. Abou-Ismail, U. A. Are the effects of enrichment due to the presence of multiple items or a particular item in the cages of laboratory rat? *Applied Animal Behaviour Science* **134**, 72–82 (2011).

57. Abou-Ismail, U. A., Burman, O. H. P., Nicol, C. J. & Mendl, M. The effects of enhancing cage complexity on the behaviour and welfare of laboratory rats. *Behavioural Processes* **85**, 172–180 (2010).

58. Abou-ismail, U. A. & Mendl, M. T. The effects of enrichment novelty versus complexity in cages of group-housed rats (*Rattus norvegicus*). *Applied Animal Behaviour Science* **180**, 130–139 (2016).

59. van der Harst, J. E., Baars, A. M. & Spruijt, B. M. Standard housed rats are more sensitive to rewards than enriched housed rats as reflected by their anticipatory behaviour. *Behavioural Brain Research* **142**, 151–156 (2003).

60. Fox, C., Merali, Z. & Harrison, C. Therapeutic and protective effect of environmental enrichment against psychogenic and neurogenic stress. *Behavioural Brain Research* **175**, 1–8 (2006).

61. Girbovan, C. & Plamondon, H. Environmental enrichment in female rodents: Considerations in the effects on behavior and biochemical markers. *Behavioural Brain Research* **253**, 178–190 (2013).

62. Simpson, J. & Kelly, J. P. The impact of environmental enrichment in laboratory rats - Behavioural and neurochemical aspects. *Behavioural Brain Research* **222**, 246–264 (2011).

63. Laviola, G., Hannan, A. J., Macrì, S., Solinas, M. & Jaber, M. Effects of enriched environment on animal models of neurodegenerative diseases and psychiatric disorders. *Neurobiology of Disease* **31**, 159–68 (2008).

64. Nithianantharajah, J. & Hannan, A. J. Enriched environments, experience-dependent plasticity and disorders of the nervous system. *Nature Reviews Neuroscience* **7**, 697–709 (2006).

65. Hermes, G. L. et al. Social isolation dysregulates endocrine and behavioral stress while increasing malignant burden of spontaneous mammary tumors. *Proceedings of the National Academy of Sciences of the United States of America* **106**, 22393–22398 (2009).

66. Garner, J. P. Stereotypies and other abnormal repetitive behaviors: Potential impact on validity, reliability, and replicability of scientific outcomes. *ILAR Journal* **46**, 106–117 (2005).

67. Bailoo, J. D. et al. Evaluation of the effects of space allowance on measures of animal welfare in laboratory mice. *Scientific Reports* **8**, 713 (2018).

68. Chamove, A. S. Cage design reduces emotionality in mice. *Laboratory Animals* **23**, 215–219 (1989).

69. Tallent, B. R., Saber, M., Law, M. & Lifshitz, J. Impact of partial cage division on aggression and behavior on long-term housing in co-housed male C57bl/6 mice. In *Abstracts of Scientific Presentations*. 2018 AALAS National Meeting 616 (2018).

70. Tallent, B. R., Law, L. M., Rowe, R. K. & Lifshitz, J. Partial cage division significantly reduces aggressive behavior in male laboratory mice. *Laboratory Animals* **52**, 384–393 (2018).

71. Anzaldo, A. J. et al. Behavioral evaluation of spatially enhanced caging for laboratory rats at high density. *Contemporary Topics in Laboratory Animal Science* **34**, 56–60 (1995).

72. Burn, C. C. Bestial boredom: A biological perspective on animal boredom and suggestions for its scientific investigation. *Animal Behaviour* **130**, 141–151 (2017).

73. Meagher, R. K. Is boredom an animal welfare concern? *Animal Welfare* **28**, 21–32 (2019).

74. Newberry, R. C. Environmental enrichment: Increasing the biological relevance of captive environments. *Applied Animal Behaviour Science* **44**, 229–243 (1995).

75. Fares, R. P. et al. Standardized environmental enrichment supports enhanced brain plasticity in healthy rats and prevents cognitive impairment in epileptic rats. *PLoS ONE* **8**, e53888 (2013).

76. Abou-Ismail, U. A. & Mahboub, H. D. The effects of enriching laboratory cages using various physical structures on multiple measures of welfare in singly-housed rats. *Laboratory Animals* **45**, 145–153 (2011).

77. Blom, H. J., Van Tintelen, G., Van Vorstenbosch, C. J., Baumans, V. & Beynen, A. C. Preferences of mice and rats for types of bedding material. *Laboratory Animals* **30**, 234–244 (1996).

78. van der Harst, J. E., Fermont, P. C. J., Bilstra, A. E. & Spruijt, B. M. Access to enriched housing is rewarding to rats as reflected by their anticipatory behaviour. *Animal Behaviour* **66**, 493–504 (2003).

79. Baumans, V. Environmental enrichment for laboratory rodents and rabbits. *ILAR Journal* **46**, 162–170 (2005).

80. NC3Rs. IAT Congress 2017 workshop summary: Playtime for Rats. (2017). Available at: https://www.nc3rs.org.uk/iat-congress-2017-workshop-summary-playtime-rats. (Accessed 20 March 2019)

81. Hawkins, P. et al. Report of the 2000 RSPCA/UFAW Rodent Welfare Group meeting. *Animal Technology* **52**, 29–38 (2001).

82. de Boer, S. F., Koopmans, S. J., Slangen, J. L. & van der Gugten, J. Plasma catecholamine, corticosterone and glucose responses to repeated stress in rats: Effect of interstressor interval length. *Physiology & Behavior* **47**, 1117–1124 (1990).

83. Armario, A., Montero, J. L. & Balasch, J. Sensitivity of corticosterone and some metabolic variables to graded levels of low intensity stresses in adult male rats. *Physiology & Behavior* **37**, 559–561 (1986).

84. Baturaite, Z. et al. Comparison of and habituation to four common methods of handling and lifting of rats with cardiovascular telemetry. *Scandinavian Journal of Laboratory Animal Sciences* **32**, 137–148 (2005).

85. Sharp, J. L., Zammit, T. G., Azar, T. A. & Lawson, D. M. Stress-like responses to common procedures in male rats housed alone or with other rats. *Contemporary Topics in Laboratory Animal Science* **41**, 8–14 (2002).

86. Sharp, J., Zammit, T., Azar, T. & Lawson, D. Stress-like responses to common procedures in individually and group-housed female rats. *Contemporary Topics in Laboratory Animal Science* **12**, 9–18 (2003).

87. Raineki, C., Lucion, A. B. & Weinberg, J. Neonatal handling: An overview of the positive and negative effects. *Developmental Psychobiology* **56**, 1613–1625 (2014).

88. Siviy, S. M. Effects of neonatal handling on play and anxiety in F344 and Lewis rats. *Developmental Psychobiology* **60**, 458–467 (2018).

89. Reis, A. R. et al. Neonatal handling alters the structure of maternal behavior and affects mother-pup bonding. *Behavioural Brain Research* **265**, 216–228 (2014).

90. Denenberg, V. H. Commentary: Is maternal stimulation the mediator of the handling effect in infancy? *Developmental Psychobiology* **34**, 1–3 (1999).

91. Maurer, B. M., Döring, D., Scheipl, F., Küchenhoff, H. & Erhard, M. H. Effects of a gentling programme on the behaviour of laboratory rats towards humans. *Applied Animal Behaviour Science* **114**, 554–571 (2008).

92. Costa, R., Tamascia, M. L., Nogueira, M. D., Casarini, D. E. & Marcondes, F. K. Handling of adolescent rats improves learning and memory and decreases anxiety. *Journal of the American Association for Laboratory Animal* **51**, 548–553 (2012).

93. Cloutier, S., LaFollette, M. R., Gaskill, B. N., Panksepp, J. & Newberry, R. C. Tickling, a technique for inducing positive affect when handling rats. *Journal of Visualized Experiments* **135**, e57190 (2018).

94. LaFollette, M. R., O'Haire, M. E., Cloutier, S., Blankenberger, W. B. & Gaskill, B. N. Rat tickling: A systematic review of applications, outcomes, and moderators. *PLoS ONE* **12**, e0175320 (2017).

95. Cloutier, S., Wahl, K., Baker, C. & Newberry, R. C. The social buffering effect of playful handling on responses to repeated intraperitoneal injections in laboratory rats. *Journal of the American Association for Laboratory Animal* **53**, 161–166 (2014).

96. Cloutier, S., Wahl, K. L., Panksepp, J. & Newberry, R. C. Playful handling of laboratory rats is more beneficial when applied before than after routine injections. *Applied Animal Behaviour Science* **164**, 81–90 (2015).

97. Bombali, V. et al. Odour conditioning of positive affective states: Rats can learn to associate an odour with being tickled. *PLoS ONE* **14**, e0212829 (2019).

98. Burgdorf J. & Panksepp, J. Tickling induces reward in adolescent rats. *Physiology & Behavior* **72**, 167–173 (2001).

99. Rygula, R., Pluta, H. & Popik, P. Laughing rats are optimistic. *PLoS ONE* **7**, e51959 (2012).

100. Mendl, M., Burman, O. H. P., Parker, R. M. A. & Paul, E. S. Cognitive bias as an indicator of animal emotion and welfare: Emerging evidence and underlying mechanisms. *Applied Animal Behaviour Science* **118**, 161–181 (2009).

101. LaFollette, M. R., O'Haire, M. E., Cloutier, S. & Gaskill, B. N. Practical rat tickling: Determining an efficient and effective dosage of heterospecific play. *Applied Animal Behaviour Science* **208**, 82–91 (2018).

102. Steenbeek, H. & van Geert, P. A dynamic systems model of dyadic interaction during play of two children. *European Journal of Developmental Psychology* **2**, 105–145 (2005).

103. Reinhold, A. S., Sanguinetti-Scheck, J. I., Hartmann, K. & Brecht, M. Behavioral and neural correlates of hide-and-seek in rats. *Science* **365**, 1180–1183 (2019).

104. Makowska, I. J. & Weary, D. M. Assessing the emotions of laboratory rats. *Applied Animal Behaviour Science* **148**, 1–12 (2013).

105. Soltysik, S. & Jelen, P. In rats, sighs correlate with relief. *Physiology & Behavior* **85**, 598–602 (2005).

106. Bartal, I. B.-A. & Mason, P. Helping behavior in rats. In Meyza, K. Z. & Knapska, E. (Eds.), *Neuronal Correlates of Empathy: From Rodent to Human* (London: Elsevier Inc., 2018), pp. 151–160.
107. Carvalheiro, J. et al. Helping behavior in rats (*Rattus norvegicus*) when an escape alternative is present. *Journal of Comparative Psychology* March **21**, Advance online publication (2019).
108. Sato, N., Tan, L. & Okada, M. Rats demonstrate helping behavior toward a soaked conspecific. *Animal Cognition* **18**, 1039–1047 (2015).
109. Li, Z. et al. Social interaction with a cagemate in pain facilitates subsequent spinal nociception via activation of the medial prefrontal cortex in rats. *Pain* **155**, 1253–1261 (2014).

10 The Rabbit

Sarah Thurston and Jan Lund Ottesen

CONTENTS

GENERAL BACKGROUND

The laboratory rabbit is one of the most widely used and yet least behaviorally understood species in biomedical research. As of 2017, it was the second most utilized United States Department of Agriculture (USDA) regulated laboratory animal species in the United States with 145,841 rabbits,[1] while in the European Union (EU), 358,213 rabbits were utilized in research in 2011.[2] The rabbit breed most used in research is the New Zealand White (NZW), *Oryctolagus cuniculus*, a descendent of the European wild rabbit[3] valued for their docile nature and greater health status compared with other breeds.[4] Another breed commonly used is the Dutch Belted which is prized in research due to their diminutive size and pigmented eyes,[5] however they tend to be more agitated in a laboratory setting and some studies have shown them to be aggressive towards caretakers.[6] Less commonly used breeds include the Half Lop, Chinchilla Bastard, Himalayan, and multiple coronary models of the Watanabe rabbit.[3] While the Watanabe model is valued for its translational medicine potential within atherosclerosis and lipoprotein metabolism research, it can pose significant husbandry challenges as it is a delicate model that must be handled with care due to their abnormally high cholesterol levels and propensity for spontaneous heart attacks.[7] Breed must be taken into account during various common husbandry procedures such as room setup and pairing. For example, Watanabe rabbits should not be housed in rooms where other rabbit pairs are often engaging in aggressive dominance displays while establishing hierarchy relationships as it may elevate their stress levels to dangerous extremes and Dutch Belted rabbits need to have extra care during social introduction pairings because they are typically in an anxious state when first placed in a cage after the shipping process and can unintentionally injure their cage mate or themselves and derail the pairing process. As with all animals, caretakers should have the proper training and knowledge necessary to provide the additional care that is required.

Multiple studies have shown that rabbits living in a laboratory setting have conserved the majority of the behavioral repertoire of their wild counterparts when provided the opportunity to express species-specific behaviors,[3,8–10] so evaluating the behavioral patterns of wild colonies of European rabbits can offer insight into the needs of domesticated rabbits. One of the key differences seen between laboratory rabbits and their wild counterparts is the amount of time spent inactive.[11] Individually housed laboratory rabbits in cage settings were found to spend the majority of their time budget in inactive behaviors (55.7%),[12] while wild rabbits spent only 33% of their time above ground in inactive behaviors.[13] This is an important difference as laboratory rabbits are prone to gastric stasis,[4] osteoporosis,[14] and other skeletal abnormalities[11,14,15] due to inactivity and restricted enclosure size, and rabbits housed in cages are prone to underdevelopment of bone tissue which can lead to fractures.[15] Due to the hazards associated with inactivity, rabbits should be encouraged to be active through species-appropriate enrichment, increased cage size or floor/pen housing when available and supervised playtimes in a larger enclosure as often as possible (Figure 10.1). When offering larger housing opportunities or playtimes in an enclosure, it is vital that they are still properly enriched to meet the animal's species-specific needs as the complexity of the space is still very significant, even in a larger enclosure.

The rabbit is a prey species[3] with a strong aversion to predators and a naturally cautious temperament[10] which can often make it difficult to determine and quantify pain behaviors in captive rabbit populations, as they are innately driven to mask painful behaviors which may mark them as targets for predation. Because of this, alternative methods have been developed to help caretakers identify when rabbits are in pain through grimace scales, which detail minor facial action units displayed when painful.[16–18] Facial grimace scale observations of the eyes, whiskers, cheeks, and nose along with a health assessment of the rabbit's normal daily behaviors can provide considerable insight into the actual post-procedural pain score of the animal. Rabbits are also nocturnal,[3,12] which means that their greatest periods of activity are at times when caretaking staff are not present in the animal facility, which can make it harder to identify when rabbits are in discomfort. Due to this strong pain-masking motivation, it is especially vital for caretakers to take the time to understand typical rabbit behavior and group dynamics so that they can be aware of behaviors that may be indicative of an underlying problem.

Another example of the importance of caretakers understanding rabbit behavior can be seen in thumping behavior. Rabbits are prone to loud thumps of their hind feet, but thumping is a behavior that they will use for multiple types of communication. It is commonly used as an alert to others of a potential predator,[3] as it is used in the wild. But thumping can also be used as a display of dominance or a display or submission depending on the accompanying behaviors.[25] It can also be utilized by the rabbits to show excitement and non-fear related behaviors such as when the husbandry staff open the food or hay containers, so it is vital that caretakers understand the subtleties behind typical rabbit behavior.

FIGURE 10.1   Supervised playtime in an open, inexpensive play area. Photo by Novo Nordisk.

ENVIRONMENT

The environment that the laboratory rabbit is housed in plays a considerable role in the physical, cognitive, and emotional welfare of the animal and thus it should be highly considered when designing a study. Environment encompasses physical housing type, social situation, and opportunities for enrichment stimulation. The types of physical housing most commonly utilized for laboratory rabbits are caging and floor pens[3] (Figure 10.2).

In recent years there has been a considerable push to modernize rabbit housing to more appropriate sizes and design with US regulations now requiring caged rabbits that weigh up to 5.4 kg have a minimum of 0.37 m² floor area per rabbit and a minimum height of 40.5 cm[19] and EU regulations requiring caged rabbits that weigh up to 5 kg have a minimum of 0.42 m² floor area for one or two socially harmonious rabbits and a minimum height of 45.0 cm (with an additional 0.3 m² per rabbit for the third to sixth rabbit).[3] In the EU, a raised platform must also be provided for rabbits housed in cages.[3] While these are the minimum standards, most caging companies offer cages that exceed these standards and allow for two or more cages to be attached to provide even more space (Figure 10.3).

While caging is the classic housing style for laboratory rabbit housing, many institutions have moved towards floor pens in recent years with regulatory guidelines in Australia and Europe

FIGURE 10.2   The two most common types of rabbit housing are caging (A) and pens (B). Photo A by Sarah Thurston, University of Michigan. Photo B by Novo Nordisk.

FIGURE 10.3   An example of a standard caging set up for paired rabbits. Two cages are attached with a removable divider and each cage contains a raised platform, water source, and food source. Photo by Austin Thomason, Michigan Photography, University of Michigan.

recommending this housing style whenever possible.[3] Floor pens allow for more species-typical expressions of behavior as they are not as limited by the four walls of a cage and can improve physical fitness by allowing for more room to fully hop, run, jump, and make quick direction changes and stretch than a cage. They provide more opportunity for social interaction and the important establishment of a hierarchic group (behaviors such as allogrooming, snuggling against each other, olfactory control, chase, biting, fighting, etc. are allowed).[20] The larger size of pen housing also typically allows for group housing, which provides additional opportunities for socialization to further increase welfare. An example of pen housing is the rabbit pens used at one of the author's institutions (Figure 10.4). They measure 2 by 1.5 m and rabbits are housed in groups of up to ten for short-term or up to eight for long-term projects. The pens are fixed to the wall at waist height to minimize bending and lifting by animal care staff, to facilitate handling of the rabbits and to ease in cleaning the floor under the pens. Each pen is individually ventilated and surrounded by plastic sheets that both discourage the rabbits from escaping and reduce room allergen levels. Each pen is divided in two along its length by a partition that has pop holes to allow the rabbits to run through and thereby give subordinate rabbits a possibility to escape from dominating rabbits. A purpose-built shelter is placed on each side of the partition, so that rabbits can hide or hop on top. The floor is solid, non-slip, and covered with a thick layer of bedding material. Natural light enters through windows and is supplemented by an artificial light cycle of 12:12 hours.

Though the size of the enclosure plays a significant role in the welfare of the rabbits, particularly in their physical ability to stretch, hop, and sit fully upright, the design of the enclosure must also be considered. A barren cage or pen does not offer many benefits to an animal, no matter how large it may be, so care must be taken to provide appropriate furniture such as boxes or huts that the rabbits can hide under or stand on top of.[21] These can be a permanent fixture of the environment or a moveable item that can be rotated through and sanitized or replaced at regular intervals. Most commercially available laboratory rabbit cages are equipped with attached huts that can be removed if needed. In addition to furniture, an acceptable environment also needs harmonious social groupings and environmental enrichment.

While social stimulation with appropriate conspecifics is the best form of enrichment because it is ever changing and unpredictable, environmental enrichment plays an immense role in laboratory rabbit welfare as well. Upon review of the published literature, it appears that until recently and still only on a small scale, rabbit environmental enrichment has been sorely lacking. Multiple studies relied on hay or a chew stick as the sole source of environmental enrichment, which is miserably

FIGURE 10.4   An example of pen housing. Photo by Novo Nordisk.

deficient, particularly in a species that requires high levels of cognitive stimulation. While hay has been shown to be particularly effective at reducing abnormal behaviors such as bar gnawing and excessive grooming,[21] it should be provided as a daily staple to all rabbits instead of being treated as a specialty enrichment since forage materials like hay are necessary roughage for rabbit digestion. Recall that in the wild, much of their active time budget is spent foraging, patrolling, and maintaining high levels of mental alertness,[22] thus this level of mental acuity must be matched as often as possible in the captive setting as well. One way that this mental energy can be exercised is through training as described in a later section. Another way is through appropriate environmental enrichment. Rabbits require novelty, rotation, and mental stimulation from their environmental enrichment[23] much in the same way that non-human primates do.[24] A suggested approach to rabbit enrichment is to group categories by value to the rabbit and rotate categories accordingly to maintain novelty.[25] Low-value items would include various manipulanda items such as toys, chew sticks, and cardboard that provide value but only for a short amount of time. High-value consumables are veterinary approved food treats, e.g., carrots or apples but should be provided in a manner that stimulates the rabbit to work to gain access which allows for natural foraging behaviors to occur which elevates the value of enrichment. An example of this can be to provide hay in a hay rack instead of just adding it to the pen/cage – in this way the rabbit will pull single hay straws from the rack resembling how it would have done in nature[20] (Figure 10.5).

Food treats should be varied in style between fresh, frozen, and dried and can be intermixed to provide originality to the rabbit. High-value destructible items are items which allow the rabbit to dig and tear, again stimulating natural foraging instincts. These are items such as paper bags, cardboard boxes, or tubes and they can be stuffed with various items such as hay, treats, or crinkled paper. Lastly, supplemental enrichment can be provided in addition to other enrichment categories but not used as a stand-alone enrichment group. This includes items such as music enrichment or gentle grooming from caretakers. While music enrichment has been shown to decrease stress in some colonies of rabbits,[26] grooming should be evaluated on a case by case basis as some rabbits will respond positively to the interaction and some will display stressful indicators and should not be candidates for brushing. Regular gentle handling by human caretakers can increase compliance in rabbits and improve welfare[27] so light grooming can be a positive rabbit-caretaker bonding opportunity when performed while closely watching for the rabbit's non-verbal cues of either stress or enjoyment. Innumerable combinations of all of these items can be made to maintain novelty for rabbits and should be thoughtfully considered when creating an enrichment schedule. Particular considerations to keep in mind when enriching rabbits are the natural inclinations for digging and foraging behaviors, so items such as foraging boxes are prized items for allowing the rabbits to

FIGURE 10.5 Hay provided in a hay rack (A) or hay provided in a tube (B). By providing hay in a manner that the rabbit has to work for it instead of just adding it to the pen/cage will resemble foraging in the wild and provides a more naturalistic experience. Photo A by Novo Nordisk. Photo B by Austin Thomason, Michigan Photography, University of Michigan.

express their natural behaviors. Wild rabbits will forage for food in the late afternoon and during the night so it can be considered to feed rabbits in the afternoon, and one study has shown that this can reduce abnormal behaviors.[28] Thought should also be given to providing appropriate gnawing items due to the rabbits' need to chew in order to wear down their continuously growing teeth.[21,29] The best environmental enrichment items provide the animals with a sense of control over their environment, allow for a greater behavioral repertoire, decrease stereotyped behaviors, allow the animal greater utilization of the enclosure and help the animal to handle the stressors of daily laboratory life.[20,36] If cognitive needs are not met by social stimulation and appropriate environmental enrichment, outward signs of stress often manifest as stereotyped behaviors that are indicators of poor welfare. Some of the most commonly seen stereotypies in rabbits include digging or scratching at the floor, excessive barbering (over grooming), bar biting/chewing, head swaying, and a depressed or hunched posture.[12,15] These stereotypies can often be mitigated by the addition of appropriate and increased environmental enrichment.

SOCIAL NEEDS

European rabbits are a naturally gregarious and social species, living in warrens consisting of complicated social hierarchies made up of a dominant male living with multiple females and subordinate males.[30] These complex social needs of their wild equivalents must be matched in the laboratory setting or it can lead to cognitive and emotional challenges that can result in outward displays of stereotypies and self-injurious behaviors indicative of poor welfare.[12,14,30,31] As the laboratory rabbit is a naturally social species, social deprivation can lead to impaired brain development,[32] increased time spent in abnormal behaviors,[14] weight gain,[33] and an increase in time spent inactive.[12] One study found singly housed rabbits displayed the most elevated fear levels and deficient behavioral repertoire of the groups tested (singly housed, pair housed, and group housed) which leads to the conclusion that they are highly stressed in this housing situation and suffering from poor mental welfare.[34] Preference tests have shown that even lower-ranking subordinate females exhibit preference for social housing over a solitary, barren cage[35] and females will work almost as hard for limited social contact as they will for food,[36] which demonstrates the tremendously high value placed by the animals on social contact. Females have also been shown to spend up to 79% of their time in close proximity to conspecifics in a laboratory setting,[30] with other studies observing them in close physical contact often as well.[10,14,37] Interestingly, even males when given a cage with protected social contact with another male via a perforated divider have been shown to prefer to spend the majority of their time near the other male rabbit and were more active, displayed normalized physiological rhythms and were less fearful of humans possibly due to the social buffering effects of having a "pseudo-conspecific."[38] Anecdotally, one of the author's institutions has paired sibling males over two years old in caged housing in a mixed-sex room with no history of severe wounding or aggression that are frequently observed in voluntary close proximity (Thurston, unpublished 2019). As previously shown, psychological damage can occur as a result of social deprivation, but there are also significant physiological challenges to consider that can introduce confounding factors into a study. These include increased oxidative stress,[39] increased white blood cell counts,[33] elevated levels of adiposity, hyperinsulinemia, and elevated heart rates.[39] Due to the significant mental and physical detriments that single housing places on rabbits, it should always be considered as the last resort.

These studies show that the social need for conspecifics is critical and must be met in both males and females. Regulatory guidelines reflect this as well with the Association for Assessment and Accreditation of Laboratory Animal Care (AAALAC) International, The Guide for the Care and Use of Laboratory Animals, The Office of Laboratory Animal Welfare (OLAW), and EU Directive 2010/63/EU, all noting that single housing of social species should be the exception and must be justified by social incompatibility, veterinary concerns, or scientific necessity as approved by the Institutional Care and Use Committee (IACUC) in the US or competent authority in the EU. Opportunities for limited socialization should still be provided even when full time social housing is not a possibility and can be provided in the form of perforated dividers, supervised playtime in open pens[40] (Figure 10.1), and at the absolute bare minimum, rabbits must be allowed to visualize and have olfactory access to other rabbits, though this is certainly not adequate to fully meet their highly social needs.

SOCIAL HOUSING

One of the greatest challenges that has perplexed caretakers in recent years is how to successfully pair house rabbits in a limited space setting. Due to evolving regulatory standards, there has been a strong push away from single housing, which was once considered the standard housing scheme for rabbits in research and is still used in many institutions. Rabbit caretakers have struggled to maintain a successful pair relationship post-sexual maturity without wounding taking place (especially in males) and in many cases have resorted to separation citing social incompatibility as the justification. There is a lack of peer-reviewed literature evaluating paired laboratory rabbit behavior which leads to a lack of understanding of acceptable rabbit behavior and what is normal aggression in rabbits. It is important to note that some aggression is normal and acceptable in rabbits[25,30] and extreme aggression that leads to wounding has observable markers or behaviors that precede wounding which allows caretakers to intervene before extreme wounding occurs. These behaviors, which are observable in both sexes, can include urine staining, chasing, mounting, and excessive thumping.[25] These behaviors are important to monitor, document, and treat with a regime of increased environmental enrichment, but *not* to separate the pair as these are normal communication behaviors. It is extremely important that daily caretakers understand that these behaviors are normal displays of hierarchy establishment in the wild and they only lead to problematic aggression when they are in a confined space in which the rabbits have no other source of distraction other than each other or do not have the proper space to perform submission displays which is why appropriate environmental enrichment and adequate space is so vital to pair maintenance. The dominant rabbit may require displays of submission from subordinate rabbits daily,[41,42] so it is not unusual for husbandry staff to see a dominant chasing subordinates daily. This is not a cause for concern as long as the subordinate rabbit has the possibility to flee as is required of the submissive display and the interaction concludes without wounding. This is a perfectly acceptable hierarchy establishment behavior and if human intervention breaks up these behaviors too often for fear of wounding, the dominant/subordinate relationship will not be able to be properly maintained and that is when actual wounding occurs. Displays of dominance can be exhibited in many different behaviors but are most often seen by urine staining, chasing (the subordinate must flee), mounting (the subordinate must allow), and excessive thumping (the dominant will usually thump his/her back feet loudly while the subordinate maintains a chin down posture to indicate a non-challenging state). When these behaviors are noted, though they are normal behaviors in the wild, they can lead to aggression when in the limited space setup of a captive environment, so increased enrichment should be added to their daily routine at this time.[23] Once caretakers have a better understanding of the underlying reasons behind rabbit behavior in the laboratory, they will have more success in maintaining the social housing experience for as long as experimentally possible. While understanding that characteristic behaviors are important, it is also important for caretakers to note the ages in which relationship breakdowns are most likely to occur within pairs. Additional monitoring can take place during this phase to assist the pairs during critical periods. Sexual maturity in rabbits is around 16–24 weeks,[43] however one study found that almost half of aggressive interactions began in weeks 10–20[23] when puberty is beginning. It is therefore recommended that pairs are monitored especially closely between weeks ten and 20, although pair breakdown and aggression can occur at any age.

Considering that rabbits are such a highly gregarious species, socializing them in the laboratory setting has been a surprisingly daunting task for many rabbit facilities. Since laboratory rabbits have traditionally been housed individually in cages,[3] the switch to social housing was an intimidating one for many institutions that were inexperienced in rabbit social housing. Previous versions of the

Guide stated that, "when it is appropriate and compatible with the protocol, social animals should be housed in physical contact with conspecifics."[44] When the current version was published in 2011 with an expanded emphasis on social housing and the guideline that social housing should be the default housing for all social species,[19] it changed the way that many institutions housed their rabbit populations.

The most efficacious way to create successful socially housed rabbits is to maintain a social group or pair consisting of same-sex siblings weaned together and monitor for the behavioral signs discussed previously. Unfortunately, not all facilities have the luxury of a breeding colony or the foresight from their researchers to order siblings from vendors and therefore have to create pairs from unrelated adults. At the time of publication, no peer-reviewed methods have been established to successfully pair adult unrelated males, thus at this time ordering sibling males from the vendor is the best option for successful intact male social housing. Adult unrelated females, however, have been shown to be able to be paired with other females successfully by utilizing the stress bonding of shipping, urine marking, and a well-defined monitoring and intervention process.[25]

HANDLING

Gentle handling methods are of particular importance when dealing with rabbits, who can be skittish while being handled. Rabbits also have a propensity to kick with tremendous force when not restrained properly. Such kicking can lead to injury to the caretaker and injury or even death to the rabbit as they may kick hard enough to fracture their spinal column.[45] Surveys done of pet rabbit owners showed that approximately 60% of pet rabbits struggled when being lifted by owners.[46] This leads to the conclusion that most untrained people are not lifting rabbits appropriately which can lead to rabbit stress, potential injury, poor welfare, and can significantly hinder the human-animal bond. Every person who will be in physical contact with a rabbit must be appropriately trained on proper handling methods before they are allowed to interact with the animals in order to prevent unintentional injury. Specifically, the rabbit should always be lifted and supported from underneath their hindquarters and held close to the caretaker's body, never dangling and *never* lifted only by their scruff or ears as this can cause extreme pain and damage.[46] Some caretakers prefer to tuck the rabbit's face into the crook of their arm to help soothe the rabbit while they are being held.Others prefer to hold the rabbit in an upward facing hold across the caretaker's chest similar to the way a baby would be held. As long as the rabbit is fully supported under the hindquarters in both situations with the other restraining hand firmly but gently on the scruff but not necessarily "scruffing," it is a matter of preference for the individual caretaker and the individual rabbit.

HUMAN-ANIMAL BOND

The human-animal bond is slightly more difficult to establish in rabbits than with some other highly social laboratory species such as dogs and pigs due to the aforementioned prey species status. Due to this, more research into the human-animal bond has been done in the companion pet literature sector[46] and the rabbit farming sector.[10] The translatability of this research to laboratory rabbits can be difficult, but still valuable. For example, evaluating appropriate methods of handling pet rabbits can have practical implications in the laboratory setting and can be used to help train new caretakers or laboratory staff.[46] These methods can include not cuddling the rabbit, only lifting the rabbit when necessary, and if necessary using proper lifting methods and educating the handler on all proper techniques. Utilization of these methods will help improve the human-animal bond by allowing the rabbit to feel safe and comfortable with the human caretaker. Though rabbits are naturally inclined to be cautious around humans, studies have shown that routine, gentle handling can actually reduce human-directed fear and aggression and improve rabbit welfare by the reduction of stress in both adults[27] and kits.[47] Everyday husbandry and veterinary duties can be an opportunity to strengthen the human-animal bond when gentle handling is utilized during health monitoring, such as checking tooth conditions and weighing the rabbit. In addition to gentle handling during routine procedures, training can be an excellent opportunity for the rabbit and caretaker to build a trusting bond as well. One study has shown that rabbits can discriminate between different humans[48] so a connection between a rabbit and a particular caretaker is possible. As with most individuals of the domesticated species, the human-animal bond can be a strong and profound connection when the human caretaker is willing to contribute the time and patience that the animal needs to form a meaningful and trusting relationship.

TRAINING

Training laboratory animals is most often associated with non-human primates, dogs, and more recently pigs, however, rarely used when working with rabbits. This is in spite of it being explicitly required in the European Directive that, "Establishments shall set up habituation and training programmes suitable for the animals, the procedures and length of the project,"[2] and it being suggested in the Guide that, "Habituating animals to routine husbandry or experimental procedures should be encouraged whenever possible as it may assist the animal to better cope with a captive environment by reducing stress associated with novel procedures or people." [19]

Training is used in many rabbit facilities and, as already described, can be an important part of the human-animal bonding process. Rabbits can learn to cooperate during handling and routine procedures and stress can be minimized by acclimatizing the animals to stay calm during stressful restraint procedures, for example by habituating the animals to be gently wrapped in a blanket[21] during blood collection.

Rabbits need to be habituated to the housing system before a training program can be established and as with training many other species, food rewards are the most efficient reinforcer. Anecdotally, the author's institution trains rabbits to voluntarily leave their pen to enter a trolley for weighing or transportation to a laboratory (Figure 10.6). While this is an attainable goal, it requires patience and quiet surroundings. Since rabbits are a prey species, even a well-trained rabbit will instinctually freeze or flee when startled by a sudden movement or loud noise so slow, steady, and predictable movements are essential to rabbit training.

Positive reinforcement training including target training can be employed when training rabbits (see Chapter 6 for details on animal training), however this is only possible in rabbits that have been

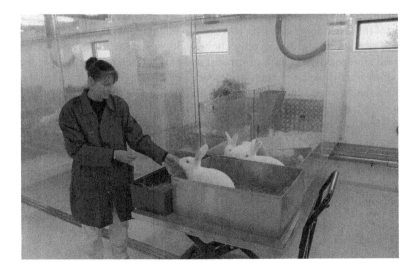

FIGURE 10.6 Rabbits trained with positive reinforcement to voluntarily leave the pens to enter a trolley for weighing or transportation to a laboratory or play area. Photo by Novo Nordisk.

habituated to their environment and previously socialized to their human handlers. Hence, the time needed to be successful is substantial. Therefore, this method is most valuable for very specific purposes and for rabbits that will stay in the facility long term.

Training will take time and resources and it requires skilled, experienced caretakers to carry out the training. Hence, commitment not only from the participating animal caretakers but also from institutional management is vital and essential if the training is to be a success. When successful, it can provide higher job satisfaction to work with animals that participate voluntarily, making it well worth the effort to invest in training if not for all rabbits then at least for some.

CONCLUSION

As understanding grows of the complex nature of the social hierarchies of laboratory rabbits and their intricate behavioral monitoring and environmental enrichment needs, many comparisons can be drawn between the care of captive rabbits and captive non-human primates. Because of these complexities and the need for a thorough understanding of behavior, it is recommended to have a behaviorist on the staff to work with all rabbit caretakers and provide behavioral interventions when necessary. If this is not feasible, it is recommended to seek input from behaviorists or personnel with experience working with other social species such as non-human primates, dogs, or pigs. This could be in the format of having workshops with a focus on rabbit behavior and where employees are allowed to come together to learn, collaborate, and share ideas. Besides having the refinement of new welfare initiatives benefitting the rabbits it can also build team cohesion, improve working climates for the employees, and decrease compassion fatigue by allowing for employee engagement in welfare initiatives.

Often rabbits are treated differently than other social laboratory species and as a collective, we need as a minimum to start holding them to the same standards that we do with other social species such as mice, pigs, dogs, and non-human primates. For example, when pigs and non-human primates have superficial wounding as a result of social housing, the social housing experience is typically maintained but when rabbits are injured as a result of social housing, they are typically separated. The cosmetic imperfections such as minor lesions or urine staining are a small price to pay for properly socialized animals.

As the field of rabbit research grows, so must our understanding of rabbit behavior and care and with that must come more effective peer-reviewed publications. The literature available in the past has offered little, contradictory, and often inaccurate insights into laboratory rabbit behavior. Some examples of this are papers advising users not to handle rabbits while others advise handling often, and enrichment studies utilizing only chew sticks or hay as the proffered enrichment item when this is hardly sufficient as a bare minimum enrichment, let alone a high value item capable of influencing study results based on their preference or lack of preference. All of these factors add to the overall idea that we need to have a better behavioral understanding of rabbits as the general lack of understanding is leading to a fear of social housing, which is a detriment to the animal's overall welfare. However, a more thorough understanding of laboratory rabbit behavior has begun to emerge.

Laboratory rabbit welfare has come so far in recent years and like for all species, still has room to improve. As long as we continue to learn and continue to have dedicated personnel at every level working to advance the daily lived experience of the laboratory rabbit, their welfare will continue to increase exponentially.

ACKNOWLEDGMENTS

Sarah Thurston would like to thank Lisa Burlingame and Jennie Lofgren as well as Patrick Lester and Tara Martin for the many years, many meetings, and many hours of dedication to building our rabbit program into something that she is so honored and proud to be a part of today! Thank you also to Austin Thomason from Michigan Photography and Amy Puffenberger for the use of Figures 10.3 and 10.5B.

Jan Lund Ottesen would like to thank the animal caretakers working with rabbits at Novo Nordisk for the commitment demonstrated in translating proposals from animal welfare experts into animal housing systems, socialization programs, and for building on their own experiences to look into new ways of working with rabbits.

REFERENCES

1. United States Department of Agriculture. *Annual Report Animal Usage by Fiscal Year*. Available at: https://www.aphis.usda.gov/animal_welfare/downloads/reports/Annual-Report-Animal-Usage-by-FY2017.pdf (2018).
2. European Commission. Seventh Report on the Statistics on the Number of Animals used for Experimental and other Scientific Purposes in the Member States of the European Union. SWD (2013) 497 final. Available at: https://eur-lex.europa.eu/legal-content/EN/TXT/PDF/?uri=CELEX:52013DC0859&from=EN
3. Lidfors, L. & Edström, T. The laboratory rabbit. In Hubrecht, R.C. & Kirkwood, J. (Eds.), *The UFAW Handbook on the Care and Management of Laboratory and Other Research Animals* (Hoboken, NY: John Wiley & Sons, 2010), pp. 399–417.
4. Mapara, M., Thomas, B. S. & Bhat, K. M. Rabbit as an animal model for experimental research. *Dental Research Journal* 9, 111–118 (2012). doi: 10.4103/1735-3327.92960
5. Spence, S. The Dutch-Belted rabbit: An alternative breed for developmental toxicity testing. *Birth Defects Research. Part B, Developmental and Reproductive Toxicology* 68, 439–448 (2003). doi: 10.1002/bdrb.10040
6. Pritt, S., Wood, J., Fetter, B. & Kline, B. *Socialization of Dutch Belted Rabbits for Use in Research: Preliminary Outcomes* (Denver, PA: World Rabbit Science Association, Castanet-Tolosan, 2008), pp. 1229–1232. Available at: http://world-rabbit-science.com/WRSA-Proceedings/Congress-2008-Verona/Papers/W-Pritt.pdf
7. The tale of the Watanabe rabbit. *Laboratory Animal* 41, 277 (2012). doi: 10.1038/laban1012-277
8. Lehmann, M. Social behaviour in young domestic rabbits under semi-natural conditions. *Applied Animal Behaviour Science* 32, 269–292 (1991).
9. Selzer, D. & St, H. Comparative investigations on behaviour of wild and domestic rabbits in the nestbox. *World Rabbit Science* 11, 13–21 (2010).
10. Trocino, A. & Xiccato, G. Animal welfare in reared rabbits: A review with emphasis on housing systems. *World Rabbit Science* 14, 77–93 (2010).
11. Dixon, L. M., Hardiman, J. R. & Cooper, J. J. The effects of spatial restriction on the behavior of rabbits (*Oryctolagus cuniculus*). *Journal of Veterinary Behavior* 5, 302–308 (2010).
12. Gunn, D. & Morton, D. B. Inventory of the behaviour of New Zealand White rabbits in laboratory cages. *Applied Animal Behaviour Science* 45, 277–292 (1995).
13. Gibb, J. A. Sociality, time and space in a sparse population of rabbits (*Oryctolagus cuniculus*). *Journal of Zoology* 229, 581–607 (1993) doi: 10.1111/j.1469-7998.1993.tb02658.x
14. Chu, L.-r., Garner, J. P. & Mench, J. A. A behavioral comparison of New Zealand White rabbits (*Oryctolagus cuniculus*) housed individually or in pairs in conventional laboratory cages. *Applied Animal Behaviour Science* 85, 121–139 (2004).
15. Morton, D. B. et al. Refinements in rabbit husbandry. Second report of the BVAAWF/FRAME/RSPCA/UFAW Joint Working Group on Refinement. British Veterinary Association Animal Welfare Foundation. Fund for the Replacement of Animals in Medical Experiments. Royal Society for the Prevention of Cruelty to Animals. Universities Federation for Animal Welfare. *Laboratory Animals* 27, 301–329 (1993).
16. Hampshire, V. & Robertson, S. Using the facial grimace scale to evaluate rabbit wellness in post-procedural monitoring. *Lab Animal* 44, 259 (2015). doi: 10.1038/laban.806
17. Keating, S. C., Thomas, A. A., Flecknell, P. A. & Leach, M. C. Evaluation of EMLA cream for preventing pain during tattooing of rabbits: Changes in physiological, behavioural and facial expression responses. *PLoS ONE* 7, e44437 (2012). doi: 10.1371/journal.pone.0044437
18. Leach, M. *Rabbit Grimace Scale (RbtGS) Manual*. National Centre for the Replacement Refinement and Reduction of Animals in Research *(NC3Rs)*. Available at: https://www.nc3rs.org.uk/sites/default/files/documents/RbtGS%20Manual.pdf

19. Council, N. R. *Guide for the Care and Use of Laboratory Animals: Eighth Edition* (The National Academies Press, 2011). Available at: https://grants.nih.gov/grants/olaw/guide-for-the-care-and-use-of-laboratory-animals.pdf

20. Ottesen, J., Weber, A., Gürtler, H. & Friis M. L. New housing conditions: Improving the welfare of experimental animals. *Alternatives to Laboratory Animals* **32**(Suppl 1B), 397–404 (2004).

21. Hawkins, P. et al. *Refining Rabbit Care: A Resource for Those Working With Rabbits in Research* (Horsham, UK: RSPCA; and Wheathampstead, UK: UFAW, 2008). Available at: https://www.research gate.net/publication/236035967_Refining_rabbit_care_A_resource_for_those_working_with_rabbits _in_research

22. Mykytowycz, R. & Rowley, I. Continuous observations of the activity of the wild rabbit, *Oryctolagus cuniculus* (L.), during 24 hour periods. *CSIRO Wildlife Research* **3**, 26–31 (1958). doi: 10.1071/ CWR9580026

23. Thurston, S., Burlingame, L. & Lofgren, J. Troubleshooting aggressive behaviors in pair housed rabbits using environmental enrichment. Abstracts of scientific presentations: 2015 AALAS National Meeting Phoenix, Arizona. *Journal of the American Association for Laboratory Animal Science* **54**, 568–668 (2015).

24. Schub, T. & Eisenstein, M. Enrichment devices for nonhuman primates. *Laboratory Animal* **32**, 37 (2003). doi: 10.1038/laban1103-37.

25. Thurston, S., Burlingame, L., Lester, P. A., Lofgren, J. Methods of pairing and pair maintenance of New Zealand white rabbits (*Oryctolagus cuniculus*) via behavioral ethogram, monitoring, and interventions. *Journal of Visualized Experiments* **133**, e57267 (2018). doi: 10.3791/57267

26. Peveler, J. L. & Hickman, D. L. Effects of music enrichment on individually housed male New Zealand white rabbits. *Journal of the American Association for Laboratory Animal Science* **57**, 695–697 (2018). doi: 10.30802/aalas-jaalas-17-000153

27. Swennes, A. G., Alworth, L. C., Harvey, S. B., Jones, C. A., King, C. S. & Crowell-Davis, S. L. Human handling promotes compliant behavior in adult laboratory rabbits. *Journal of the American Association for Laboratory Animal Science* **50**, 41–45 (2011)

28. Krohn, T. C., Ritskes-Hoitinga, J. & Svendsen, P. The effects of feeding and housing on the behaviour of the laboratory rabbit. *Laboratory Animals* **33**, 101–107 (1999). doi: 10.1258/002367799780578327

29. Poggiagliolmi, S., Crowell-Davis, S. L., Alworth, L. C. & Harvey, S. B. Environmental enrichment of New Zealand White rabbits living in laboratory cages. *Journal of Veterinary Behavior* **6**, 343–350 (2011).

30. DiVincenti, L., Jr. & Rehrig, A. N. The social nature of European rabbits (*Oryctolagus cuniculus*). *Journal of the American Association for Laboratory Animal Science* **55**, 729–736 (2016)

31. Podberscek, A. L., Blackshaw, J. K. & Beattie, A. W. The behaviour of group penned and individually caged laboratory rabbits. *Applied Animal Behaviour Science* **28**, 353–363 (1991)

32. Baumans, V. Environmental enrichment for laboratory rodents and rabbits: Requirements of rodents, rabbits, and research. *ILAR Journal* **46**, 162–170 (2005)

33. Fuentes, G. C. & Newgren, J. Physiology and clinical pathology of laboratory New Zealand white rabbits housed individually and in groups. *Journal of the American Association for Laboratory Animal Science* **47**, 35–38 (2008).

34. Trocino, A., Majolini, D., Tazzoli, M., Filiou, E. & Xiccato, G. Housing of growing rabbits in individual, bicellular and collective cages: Fear level and behavioural patterns. *Animal: An International Journal of Animal Bioscience* **7**, 633–639 (2013). doi: 10.1017/s1751731112002029

35. Held, S., Turner, R. & Wooton, R. Choices of laboratory rabbits for individual or group-housing. *Applied Animal Behaviour Science* **46**, 81–91 (1995).

36. Seaman, S. C., Waran, N. K., Mason, G. & D'Eath, R. B. Animal economics: Assessing the motivation of female laboratory rabbits to reach a platform, social contact and food. *Animal Behaviour* **75**, 31–42 (2008).

37. Huls, W. L., Brooks, D. L. & Bean-Knudsen, D. Response of adult New Zealand white rabbits to enrichment objects and paired housing. *Laboratory Animal Science* **41**, 609–612 (1991).

38. Lofgren, J. L. et al. Innovative Social Rabbit Housing. 2010 AALAS National Meeting Atlanta, Georgia. *Journal of the American Association for Laboratory Animal Science* **49**, 659 (2010).

39. Noller, C. M. et al. The influence of social environment on endocrine, cardiovascular and tissue responses in the rabbit. *International Journal of Psychophysiology* **88**, 282–288 (2013). doi: 10.1016/j. ijpsycho.2012.04.008

40. Garcia-Gonzalez, R. et al. A Company's Global Efforts to Refine Enrichment: Exercise Pens and Group-Housing for Laboratory Rabbits. Abstracts of Scientific Papers 2017 AALAS National Meeting. *Journal of the American Association for Laboratory Animal Science* **56**, 574–694 (2017).
41. Mykytowycz, R. Territorial marking by rabbits. *Scientific American* **218**, 116–126 (1968).
42. Lockley, R. M. Social structure and stress in the rabbit warren. *Journal of Animal Ecology* **30**, 385–423 (1961). doi: 10.2307/2305
43. River, C., Pritchett-Corning, K. R. & Laboratories, C. R. *Handbook of Clinical Signs in Rodents and Rabbits* (London, UK: Charles River Laboratories International, Incorporated, 2011).
44. Clark, J. D., Gebhart, G. F., Gonder, J. C., Keeling, M. E. & Kohn, D. F. The 1996 Guide for the care and use of laboratory animals. *ILAR Journal* **38**, 41–48 (1997). doi: 10.1093/ilar.38.1.41
45. Flecknell, P. Restraint, anaesthesia and treatment of children's pets. *In Practice* **5**, 85–95 (1983).
46. Bradbury, A. G. & Dickens, G. J. Appropriate handling of pet rabbits: A literature review. *The Journal of Small Animal Practice* **57**, 503–509 (2016). doi: 10.1111/jsap.12549
47. Anderson, C. O., Denenberg, V. H. & Zarrow, M. X. Effects of handling and social isolation upon the rabbit's behaviour. *Behaviour* **43**, 165–175 (1972).
48. Davis, H. & Gibson, J. A. Can rabbits tell humans apart?: Discrimination of individual humans and its implications for animal research. *Comparative Medicine* **50**, 483–485 (2000).

11 The Dog

Carolyn Allen, Dorte Bratbo Sørensen, and Jan Lund Ottesen

CONTENTS

GENERAL BACKGROUND

Dogs have evolved alongside man, and even though researchers do not agree entirely on how, where, and when domestication started, it is fair to say that the dog has been bred by humans for thousands of years – and longer than any other species of domesticated animal.[1–3] Dogs have been bred not only as companions, but also to suit human needs and preferences. The desire for humans to have a dog as a protector, companion, hunter, retriever, and to assist with special needs, has resulted in hundreds of breeds and further variations of those.[1] Dogs have also been used in biomedical research since – and perhaps before – the days of Galen of Pergamon (AD 125–210).

As of 2017, the United States Department of Agriculture (USDA) reported that 64,707 dogs were used,[4] while in the European Union (EU), 17,896 dogs were utilized in research in 2011 (latest available official figures).[5] Most dogs used for experimental procedures are purpose bred, and in the EU purpose breeding of laboratory dogs is a legal requirement.

Purpose-bred dogs are generally conditioned to handling for study as well as husbandry procedures. This habituation and socialization begin with the vendor and is more or less dependent on client request. Beagles tend to be the most common breed used and they are known to be a friendly, tolerant, and social breed of dog to work with in a laboratory setting.

Incorporating the 3Rs into a laboratory dog enrichment and training program is the expectation and legal requirement according to the EU directive. The process of refinement around housing dogs (short-term as well as long-term), should be a constantly ongoing action to continuously improve the overall animal welfare of the dogs. Refinement in all its aspects should also be emphasized; for example, when describing/presenting programs to the industry or academia facility tours should be offered and knowledge on dogs used for experimental purposes should be disseminated to students and the public (e.g., www.comeseeourworld.org and www.novonordisk.com/research-and-develop ment/bioethics/animal-ethics/responsible-use-of-animals.html).

In this chapter we will address the importance of frequently evaluating individual dog behavior, compatibility of group-housed dogs, and the caging/housing style, as well as the effectiveness of enrichment, training, and rehoming programs. Such ongoing evaluation is important to identify and limit variables that can lead to reduced welfare.

DOG SENSES AND COGNITION

Dogs are social animals evolved to live in multispecies societies and today they often have more interactions with humans than with other dogs.[6] Dogs are therefore able to communicate with both humans and dogs. The sensory capabilities of dogs, however, are different from those of humans and we need to be aware of these differences when designing housing facilities and enrichment for dogs. Possibly due to the close bonds that may often arise between dogs and the humans that care for them, humans tend to forget that the sensory input from the surrounding world is quite different for a dog compared to that of humans.

Dogs' visual abilities are different from those of humans. Even though dogs, like humans, have functional vision both during day and night, dogs are, most likely, better adapted to dim light. In contrast, their visual acuity (the clarity or sharpness of vision) is considered to be worse than that of humans.[7] Moreover, dogs have dichromatic vision with only two photoreceptors in contrast to humans, which are trichromatic. It is not fully established how dogs perceive colors – and if they can perceive ultraviolet light. It has been suggested that dogs may be unable to perceive differences between, e.g., green and gray and that they perceive red as black.[6] However, it still remains to be fully established which colors dogs do see and how the light conditions affect their color vision.[7] It is important to note that dogs are highly sensitive to visual signals such as pointing from humans and also eye contact seems to be important in human-dog communication – and possibly even more important than auditory cues such as calling the dog's name.[7,8] When training dogs it is vital to remember that dogs will take any visual cues that can assist the dog in solving the task – both the signals we consciously provide (such as a hand signal or presentation of a target) and those we give unknowingly (e.g., facial expressions or movement of the upper body while giving a hand signal).

Dogs' hearing abilities also differ from that of humans. Dogs and humans have the same lower limit in hearing range, whereas dogs are able to perceive ultrasound (sounds above 20 kHz and not audible for humans) up to approximately 45 kHz. Dogs are stated to be able to hear sounds up to four times quieter than humans and they are highly sensitive, to e.g., barking.[9] Dogs' hearing abilities are developed for hunting and they are highly skilled at pinpointing the origin of a sound.[6] All in all, it is fair to assume that dogs are much more sensitive to auditory stimuli than humans are.

Another important characteristic to consider when working with dogs is their excellent olfactory sense.[10] Dogs are known to be able to detect the scents of, e.g., drugs, explosives, and flammable materials;[11] they have been used in diagnosis of different diseases[12] and they have been used by law enforcement agencies to detect human scents of live and deceased bodies.[11] This is important to recognize when evaluating, implementing, and managing housing type (e.g., housing males in close proximity to females in heat), meaningful enrichment, and activities that are most rewarding for particular breeds of dog. In domestic dogs, regular nose work will make the dogs more optimistic and most likely increase their welfare.[13] It is a reasonable assumption that laboratory dogs will benefit from olfactory enrichment in the form of nose work in the same way. For example, beagles may benefit from having an opportunity to use their sense of smell and desire to track to find, e.g., hidden meat-flavored treats throughout an open area and being allowed to find each treat on their own accord.

While it is important to take into consideration what behaviors are natural for dogs in general, such as the desire to track/sniff, a large part of canine behavior repertoire that can sometimes be taken for granted in utilizing this species as a research animal is the incredible and undeniable human-animal bond between dogs and humans.

Dogs have an inherent ability to communicate and cooperate with humans. Interestingly, both wolves and dogs can cooperate with humans in a string-pulling task; but the dog tends to wait for the human to take the initiative, whereas the wolf, to a higher extent, initiates action.[14] When it comes to understanding human signals (e.g., indicating the location of hidden food) dogs have been found to be more skillful than great apes.[15] This suggests that domestication of dogs has enabled them to communicate with humans in unique ways.[16–18]

Dogs are a well-known social species, creating bonds, relationships, understanding of social status, and hierarchies with other dogs and generally preferring to be social, which also includes with humans. Dogs are one of the few species that seek out eye contact with humans,[15,17] and there are indications that dogs can have "emotional understanding," e.g., dogs approach a human more if the person is crying than if the person is humming, and dogs are more responsive to crying babies compared to babbling babies.[19,20] It has also been reported that dogs show complex forms of dog-human communication which under similar procedures cannot be demonstrated in wolves.[17]

Attachments to the dogs we work with are quite common and encouraged. Subjectively speaking, positive human interactions are the absolute best type of enrichment we can offer to laboratory dogs. While many facilities may house large numbers of dogs and conduct multiple studies each year, personal experience with creating enrichment and behavior programs that are managed in such a way in which dogs are viewed as individuals has been of monumental reward. Recognizing individual dogs' differing personalities, needs, desires, responses, fears, and demeanor has been key for a successful program(s) – this is further elaborated on later in "Treating Dogs as Individuals."

THE HOUSING ENVIRONMENT

The laboratory housing environment for dogs can vary from country to country and from stainless steel to caging racks or wall-hung cages, kennel/run type enclosures, open room concept, and access to play rooms and/or outdoor spaces.

In the US, the Guide for the Care and Use of Laboratory Animals (the Guide)[21] requires a minimum of 0.74 m^2 floor area per dog less than 15 kg, minimum 1.2 m^2 floor area per dog up to 30 kg and minimum 2.4 m^2 floor area per dog more than 30 kg and the cage height should be sufficient for the animals to comfortably stand erect with their feet on the floor. However, the Guide also states that enclosures that allow greater freedom of movement and unrestricted height (i.e., pens, runs, or kennels) are preferable. Housing in indoor pens also gives the possibility to enrich with platforms and ramps, which provide a choice of resting place and observation opportunities, offering the dogs visibility across the room (Figure 11.1). EU regulations[22,23] require that dogs weighing up to 20 kg to have a minimum of 4 m^2 floor area for one or two dogs and a minimum height of two meters (with an additional 2 m^2 per dog for each additional dog). For dogs weighing more than 20 kg a minimum of 8 m^2 floor area for one or two dogs is required (with an additional 4 m^2 per dog for each additional dog). Furthermore, the EU regulations require that dogs shall, where possible, be provided with outside runs and that dogs shall not be single housed for more than four hours at a time.

Aside from space regulations and requirements for dogs, the cage or enclosure environment should ideally offer enough space for the dog to explore and eliminate in separate areas from where she/he sleeps. Visual and possibly tactile access to neighboring compatible dogs is important when dogs are required to be separated for study or husbandry related activities, as are visual barriers that offer privacy if so desired.

An aspect of housing that should be considered is provision of visual enhancements such as elevated resting boards or surfaces to climb. If the dogs are housed in pens, an option is to provide raised platforms large enough for several dogs (Figure 11.1). The dogs use these platforms a lot, especially when people are entering the room as the dogs are then able to see who is entering.

Also "viewing windows" or areas of caging that are popped out ("pop-out" type viewing windows) will allow dogs to see and further experience/evaluate the room around them (Figure 11.2). This affords the dog the opportunity to view her/his surroundings, room activities and other dogs within the room and provides further structural enrichment of the primary enclosure.

The "pop-out" style viewing windows in dog runs are becoming ever more popular at industry vendor trade shows, as they are subjectively known to lead to lower amounts of barking, and therefore presumably lower amounts of stress, as the dogs can see what is going on and may be able to better cope with room activities/possible stressors (Figure 11.2). There are now even commercial

FIGURE 11.1 The individual pens can be connected in a flexible way, allowing dogs to be standard-group housed, but still fed individually. Mature dogs are typically housed in harmonious groups consisting of two to four animals each. These indoor pens are enriched with platforms and ramps, which provide a choice of resting places and observation opportunities, offering the dogs visibility across the entire room. The dogs are given various other types of toys for playing and biting as well.

FIGURE 11.2 Pen-style kennel with "pop-out" viewing windows allowing the dog(s) further visualization of the entire room.

fencing companies for homes/yards that offer fence walls with bubble type pop out windows for pet dogs to visualize outside their fenced area.

The well-known loud bark or bay from a beagle or hound can become excessive and almost deafening in a large room or area containing dogs, hence the use of sound-absorbent materials in the dog facility should be considered as this will benefit both dogs and employees. As mentioned earlier, dogs possess excellent olfactory and auditory senses, not to mention the toll that barking can have on human ears over time, so limiting room and facility noise should be a consideration for establishing and maintaining a dog colony. It is preferable, though, to reduce the intensity of barking and other vocalizations by allowing the dogs to have visual contact with other dogs and possibilities to survey the room.[24] Furthermore, it is recommended to house the dogs in smaller rooms to further decrease noise level.

Single housing is not ideal for social species such as dogs, and according to EU legislation social species should – as a general rule – be housed in groups. Hence, the ability to co-house groups of compatible dogs should be afforded as a given, assuming study requirements do not require single housing. It is preferable to have large pens for group housing that can also be divided into individual pens and – if needed – connected in a flexible way (Figure 11.1). Such a pen system will allow for the dogs to be standard group housed, but, e.g., fed individually. All mature dogs are typically housed in harmonious groups consisting of two to four animals each. All the indoor pens are enriched with platforms and ramps, which provide a choice of resting place and observation opportunities, offering the dogs visibility across the room.[25]

For a species that naturally forms attachments with humans, and which is typically active, engaged, curious, playful, and social, ultimately, long-term housing in a caged or limited environment has the potential to result in decreased welfare. The limited environment will most likely result in unmet behavioral needs and preferences, including, e.g., lack of possibilities to forage and engage in appropriate social interactions such as play, and/or the ability to flee if agitated or provoked (such as by a cage mate). These limitations may result in a general lack of stimulation such as physical exercise and lead to frustration, boredom, and reduced social skills. The dogs may become easily agitated, develop stereotypic behavior, bark excessively, and/or become increasingly difficult to handle.

In the laboratory environment, dogs may also have limited control over potential stressors, i.e., stimuli that disrupt homeostasis and elicit a stress response.[26,27] Potential stressors can be, e.g., physical stressors such noise or cold, psychological stressors such as stimuli eliciting fear, or social stressors. Study activities, facility sounds, and noise from other dogs/species in close proximity may therefore be stressful for the dogs. The concept of stress is complex, and a thorough discussion is outside the scope of this chapter. The interested reader may refer to, e.g., the review by Pacak and Palkovits (2001) or *The National Research Council Committee on Recognition and Alleviation of Distress in Laboratory Animals* (2008).[26,28] Both stress and distress represent potential complications in a wide range of experiments, and should be proactively addressed by good experimental design focusing on reducing physical and mental strain on the dogs. "Doing our best and then some" should be the expectation when creating and maintaining a dog enrichment and behavioral management program. In addition to experiencing a reduction in animal welfare, a stressed animal – accounted for or unknowingly unaccounted for – may present with unwanted (and unknown) variables as a study subject.[28,29] Ultimately, paying attention to and providing alternatives to, or remediation of, known stressors should be a top goal of any program housing laboratory dogs.

In time, a barren environment and/or potentially many unalleviated stressors may lead to chronic stress[29] and the resulting development of abnormal repetitive behaviors (stereotypies) such as pacing, digging at the enclosure walls, or flooring to the point of injuring nails, or spinning in circles within the enclosure. Concurrently, self-directed behaviors such as self-mutilation and self-soothing by paw sucking have the potential to cause chronic cysts and/or granuloma sores. These coping strategies and stereotypies may have further potential to become chronic, which creates the potential to become physically, physiologically, and mentally detrimental in the long term. Chronic stress that the dog cannot relieve or properly cope with, can become a vicious cycle without attention and effort to break it.

ENRICHMENT IN THE DOG LABORATORY FACILITY

There are many categories of enrichment, and it is important to consider what types of enrichment are important to a dog. An enrichment program can then be built around those categories. The goal of an enrichment program for dogs should be to not only meet species-appropriate needs (physiological, emotional, social, and cognitive), but to strive for exceeding them. For example, provision of basic physiological needs such as food, water, exercise, and shelter are easily/legally met in the laboratory environment, but exceeding them entails doing more, such as providing novel space to encourage exploration, provide food in a foraging feeder, and implementing scheduled leash walks (Figure 11.3) and/or human interaction play programs. One way of addressing many of the above enrichments is to have adequate space for exercise and recreation (playrooms or playpens), where all dogs have regular access. The dogs should be put together in harmonious groups that allow them to experience a complex and varied environment including positive, species-specific interaction with other dogs and positive interactions with humans (see Figure 11.4). If the facility does not include a specially designed room that can be used for exercise and play, the many corridors in the facility can often easily be utilized. Ultimately, we want our dogs to be well-adjusted and to carry low burdens of stress in order to be the best study candidate/model possible. The more we can do to identify and remove or lower potential stressors the better. Knowing how to properly enrich them, especially as individuals rather than in a "herd" mentality, is vital to achieving this goal for all laboratory dogs.

Providing enrichment with a focus on individual dogs calls for "a toolbox" with several categories of enrichment and various items in each category. It is then possible to incorporate one or several items or events into each day and pay attention to how the dogs interact with and respond to each of the stimuli. It is important to note which stimuli are rewarding for which of the dogs and to make sure that the resulting activity is in fact positive and that the stimuli provided are not, e.g., causing fear or possible stress. For example, in dogs, newly arrived, noisy, or large enrichment items may be fear-eliciting and the dogs must be allowed time to habituate to all potentially frightening aspects of the new facility. The enrichment programs in the dog facility should continuously improve on enrichment strategies to prevent boredom and truly make a positive impact and difference for the daily life of each individual dog. Enrichment for dogs can be divided into several categories. Table 11.1 presents examples of enrichment inspired by Coleman and Novak (2017)[30] and the joint experiences by the authors of this chapter. Optimally, all categories should be considered when planning enrichment programs (Figure 11.5).

FIGURE 11.3 Taking dogs for a leash-walk outdoors is considered an enrichment both for private dogs and laboratory dogs. This kind of enrichment provides both a large variety of sensory input, changes in the physical environment (e.g., walking on grass, gravel, or roads), exercise and interaction with the caretaker (and other dogs, if more dogs are walked at the same time).

FIGURE 11.4   Play and exercise areas. Designated playrooms with multiple textures and structures for climbing and exploration, as well as visual and other sensory enhancements. Play and exercise areas can be indoor as well as outdoor.

HANDLING AND MANAGEMENT

An age-divided socialization program should be introduced, and this means that positive human interactions should ideally begin with the vendor and be continued as soon as dogs arrive in the facility. The early socialization program should be applied to puppies from three to ten weeks of age in collaboration with the vendor and should include, e.g., daily physical contact/gentling while cleaning and feeding, and play and close physical contact with humans both within the litter and individually. Moreover, the vendor should ensure familiarity with normal sounds and disturbances in a facility. If possible, the user of the dogs should discuss with the vendor possible facility-specific stimuli and conditions that the dogs, with advantage, could be habituated to from the very start.

Litters already gently handled from day three until day 21 have been demonstrated to be calmer later in life and this is advantageous to the emotional development and welfare of the dogs.[31] Puppies enter into a period of great change and sensitivity with regard to social relationships at approximately three weeks of age and they have a short critical period where lasting patterns of behavior are acquired and social relationships are developed. Therefore the most optimal time for puppies to become well-balanced in their relations with people and other dogs is from approximately six to eight weeks of age and before they are 16 weeks old.[32] At this age it is especially important that the puppies are familiarized with conditions likely to be encountered during subsequent use. Hence it should, e.g., be agreed with the vendor that the dogs are gently handled daily until they are received in the facility.

Dogs are received on the dock and either hand carried or placed in transport carts with visual and tactile access to other newly received dogs. Once they arrive in their housing room, hands-on assessments are performed, as well as quick friendly greetings with caregivers, and they are then

TABLE 11.1
Examples of Enrichment for Canines

Categories	Sub-Categories	Examples
Physical cage structure		Viewing windows, resting boards, visual barriers, elevated structures, areas to explore. Different types of walking/resting surfaces, raised resting areas, hard flooring vs. grid flooring.
Sensory stimulation	Olfactory	Wiping a scent on cage bars, offering a toy with a scent, utilizing calming pheromone diffusers or sprays, scented bubbles, hiding flavored treats in toys, leaving a scent trail in an open play space, etc.
	Auditive	Dog appropriate music (low bass, even toned, "calming" music), toys that make a noise, podcasts, or audiobooks (mimic human voices talking and can help to improve welfare and lower stress.
	Visual	Ability to view activity within home room via "pop-out" type windows within cage environment (Figure 11.2) or raised platform in the pens allowing dogs to see the entire room (Figure 11.1).
	Gustatory	Food enrichment such as meat, treats, dog-friendly foodstuffs presented and rotated unpredictably, and/or smeared on toys, frozen, hidden, hand fed to facilitate training, etc.
	Tactile	Frozen cubes (Figure 11.5), toys with differing textures, destructible/manipulative items such as cardboard boxes (with or without treats inside), play, and other physical contact with other individuals.
Social interaction	Same species (dog-dog)	Full-time compatible groupings with other dogs, exposure to other compatible dogs in novel spaces, play time.
	Interspecies (dog-human)	Positive interactions such as grooming, cuddling, and positive reinforcement training.
Physical exercise		Water pools, digging areas. Leash walks (Figure 11.3), exercise/play opportunities outside of the home enclosure (Figure 11.4 A–C) or on the home room open floor, searching for hidden treats, interaction with other dogs, exercise/play opportunities outside of the home enclosure.
Cognitive challenge		Puzzle boards/toys, challenging and interactive toys, forage mats, water pools, and digging areas with hidden treats. Positive reinforcement training.

Some of the categories and sub-categories may be overlapping (such as food enrichment which will stimulate vision, smell, touch, and taste – and even hearing, if it is really chewy or crunchy). Obviously, there are several other items, events, and constructions that can be employed when planning an enrichment program for laboratory dogs.

released on the room floor with several others (of the same sex) for open floor play. Alternatively, the use of flexible pens can allow for a larger group of dogs to explore the new surroundings together (Figure 11.1).

Handling dogs when they are received at the facility, talking to them, and allowing short stints of open floor play (especially after what is typically a long truck ride to get to a facility from the vendor) are extremely positive additions to the dog receipt and housing process.

During this time, it is very easy to evaluate individual dog demeanor, confidence level, playfulness, and the beginnings of a hierarchy. In this way dogs are allowed to pick their own partner(s) and the author has had great success with long term compatible groupings using this "meet and greet" method. Prior to instituting this sort of system with newly received dogs, timidity/shyness were common behavioral classifications and fights and aggression were also more common, causing new pairs/groups to be broken up early on, when social support is much needed. The ultimate goal from the moment of receipt should always be to decrease stress and manage the potential for stress as much as possible. By allowing dogs to initially form their preferred relations, some control and

FIGURE 11.5   A frozen chicken broth and pea cube provides sensory stimulation as well as food enrichment.

choice is given back to the dogs, resulting in increased welfare and decreased aggression. During the dog's stay in the facility, it should be acknowledged that dogs in a laboratory environment experience so many stimuli and have limited control over their environment and daily schedules. The people interacting with them each day make a huge difference and no interaction is inconsequential. During daily management routines, people will walk to and from the housing facility, and simply walking into a room housing many dogs can have an impact on the dogs. Dogs, in their own unique way, will often be at the front of their cages/enclosures inviting our attention, and many of us stop and pet each one of them. This may be a challenge if on a tight time schedule, but keep in mind that such positive human contact may be very beneficial for the dogs.

TRAINING

Properly socializing young dogs, not only to other dogs but also to humans is strongly recommended. The purpose of socialization is both to assure well-adjusted candidates for study or research use and a possible future for the dog as an adoption candidate. Dogs should ideally be habituated to all standard procedures they will come across during their stay in the facility. Allowing dogs to stay with their dam until they are fully weaned is critical. It is also important to keep puppies in stable social groups where hierarchies can be established and maintained as well as affording the opportunity to learn social cues for appropriate/tolerated behavior.

After the initial socialization at the vendor this should be continued with the young dogs from the day of arrival at the facility. The optimal time of arrival is at eight to ten weeks of age and the socialization should continue until maturity. If the dog is not arriving at the research facility when she/he is eight to ten weeks old, a continued socialization program at the vendor is vital. Preferably, such a program should be developed in a collaboration between vendor and user in order to ensure the best possible socialization and training on the young dogs, taking into consideration the procedures they will be subjected to in later life.

The socialization of the puppies and the young dogs should include: a) daily physical contact with humans while cleaning and feeding, b) either play with the group of dogs housed together or other close physical activities with humans five times (in the beginning) or twice (later socialization program) a week for 15–30 minutes, and c) individual playing, going for a walk, or training for the future experimental procedures at least twice a week for 15–30 minutes.[25]

Optimal ways of working with laboratory dogs include investing in quality human interaction, utilizing positive reinforcement strategies for training, and providing opportunities for the dog to have a choice and some control over his/her environment (please also refer to Chapters 5 and 6). At one facility, training of dogs has taken on a life of its own and is honestly one of the most rewarding aspects of our program. Several years ago, it became apparent that doing some basic target training may alleviate the time husbandry technicians were spending trying to get dogs to shift into their appropriate cages for feeding time or moving from the enclosures for study procedures. Investing in the first few dogs, training turned into a few more dogs which turned into a few more dogs and soon training became the expectation and a major part of our behavioral management and enrichment program. Picking the most concerning behaviors and working to remediate them is key; it is important not to train too many behaviors at once.

Training and working closely with a dog and watching him/her learn and listen to your cues is extremely rewarding and further builds not only a bond between trainer and trainee (the dog) but also offers excellent opportunities for positive human interaction, mental and physical stimulation, enrichment, choice (to participate or not), scent, and rewarding the dog (either with treats or verbal praise/positive contact such as petting, play with a valued toy, etc.).

Positive reinforcement via basic target training has worked well for our circumstances and facility. Dogs undergo target training mainly for the following circumstances, which were the most concerning behaviors we wanted to remediate:

- Shifting into the appropriate cage on cue.
- Follow a target within or from cage/pens to a designated area.
- Remain calmly in their cage when doors are opened (and not jump or flop out).
- Move away from the feed bowls when given a cue, as many dogs tend to stand in their bowls when the husbandry staff come around to remove the bowls for cleaning each day.

Prior to initiating a training program for our dogs, we found that we were unknowingly reinforcing the behaviors we truly wanted to remediate. As interaction with humans is highly valued by the dogs, and any behavior that leads to increased positive interaction with humans will be reinforced. For example, we caught the dogs as they leaped/flopped out of their cages and then cuddling, holding, laughing, etc. as the dogs were returned to their cage. In other cases, we reached into the cage and physically gently moved or picked up a dog that would not shift. Yet another example could be that we physically moved a dog from his/her food bowl by gently guiding the dog away while, e.g., talking to the dog.

Building a successful training program with dogs requires dedication and commitment. Clear and consistent direction and proper timing/reward, as well as consistency in commands and personnel performing the training and maintenance of training once the desired behavior is learned, is key!

TREATING DOGS AS INDIVIDUALS

When it comes to treating dogs as individuals, naming of animals is a hot topic in parts of the laboratory animal community. It is widely discussed whether naming the animals in our care is a good idea. There are pros and cons to naming animals on both sides of this argument, but, from the point-of-view of the authors of this chapter, the pros always tend to hit closer to home when having this discussion:

1) When training dogs it is really helpful to call them by name – they listen, they respond, they know you are focusing on them (especially when training in a room with many animals and high activity) and they learn in this manner.
2) When a facility houses many animals, bringing a specific animal up in conversation with, for example, a veterinarian, translates so much easier when referring to the dog by name. For example, "dog #1234567 hasn't been eating well," and you want to let the veterinary staff know. In a facility with hundreds of animals, dog #1234567 is not necessarily an ID that most of us will immediately recognize, but if we say, "Buddy hasn't been eating well," the majority of those involved/invested in this dog's health care are going to know exactly who we are talking about.
3) The fact that the dog has a name makes it more intuitive for the trainer to bond with the dog. The dog will more likely be perceived as an individual and not just a tool.

From the "con" side, names should never be used as the sole identification in a regulated species record. In the US animals are assigned a USDA identification number, and the name should be secondary to identifying an animal.

The possibility of bias or favoritism is a reality as well, as is the belief that giving names stirs feelings for an animal. However, one could argue that naming animals also increases the likelihood that the people caring for them are more invested in their well-being and tend to notice nuances or when things may be a bit off with a particular animal. Thus, feeling for our animals and always looking out for their best interest should be the expectation, not the exception. A caring culture towards laboratory animals requires the capacity for empathy (as elaborated in greater detail in the chapter on Culture of Care) and it seems fair to state that empathy is easier to achieve with "Buddy" than with "dog #1234567."

Regardless of assigning names or not, based on first-hand experience, strong bonds can, and should, be built with our laboratory dogs. While positive human interaction is a huge part of our programs, the time and emotions invested in bonding with dogs can take a toll, while at the same time be extremely rewarding. Aside from daily interactions and routine husbandry, exercise and enrichment activities, a few examples of further increased human interaction efforts may include (but are certainly not limited to):

• Dogs displaying stereotypic or other concerning behavior.
• Dogs that need to be single housed or exempted from social housing due to study design, health concerns, or aggressive behavior.
• Dogs receiving increased acclimation efforts for study procedures.
• Dogs receiving increased acclimation efforts for adoption preparation.
• Dogs that may experience serious side effects while on a study or during/after a procedure.

These circumstances (and more) may require increased efforts and possibly a higher output of emotion or attachment on our part, but can also lead to better care, further investment in animal welfare, understanding, and compassion towards our dogs.

Many facilities already have (or are planning) adoption strategies/programs in place that allow dogs to be placed in homes after the studies are completed. For example, at AbbVie the rehoming program has successfully placed hundreds of dogs in homes since its inception; and many of those dogs go home with staff members! We ultimately want the dogs that are released for adoption to be positive ambassadors for our program and the retired research dog community. Evaluating behavior and demeanor, as well as identifying any concerns prior to adopting dogs out, is essential. Close communication between the parties involved in adoptions/retirements is also beneficial.

Depending on the local legislation, a permission to rehome laboratory dogs may be needed. Secondly, the dogs should carefully be evaluated by trained and highly experienced behavior and animal care staff with input from on-site veterinary staff. The assessment of the dogs must include a physical examination and behavioral evaluation, further ensuring the likelihood of adapting and thriving in a new home for the rehomed dogs. The assessment must be backed up with the individual dog's known history. The focus must always be seen from the dog's perspective. If a dog is not found suitable for rehoming, a more appropriate use may need to be found for that dog. The EU legislation dictates that the new owners must be informed about the dog's history and instructed in which measures are required to ensure a successful rehoming. They should also receive information on feeding, vaccinations, advice on good dog health, and (in some countries) of mandatory insurance. If possible, access to the institution's veterinarian(s) and the animal care technicians for advice (in the EU) should be offered. Finally, it is beneficial if future owners of the rehomed dogs are able to provide written feedback after a defined period. In most cases it is very rewarding to hear how the dogs are doing in their new homes. The Laboratory Animal Science Association (LASA) has produced a very good guidance document that can provide recommendations before rehoming[33] and recommendations from the authors of "an observational test and survey of new owner study" can be found in Döring et al.[34]

CONCLUSION

Enriching laboratory dogs' lives includes enriching their environment and providing opportunities for engaging in species-appropriate activities, including positive interactions with humans. As the laboratory animal field evolves, more attention and emphasis continue to be placed on expanding species-appropriate environmental enrichment opportunities. An example is facilities that have invested in playrooms or areas for dogs to engage in increased play and exploration. Outdoor/exercise areas, once thought to be far out of reach or scope for our industry (at least in the US), are starting to spring up, and based on the authors firsthand experience, provide extraordinary opportunities for dogs to play, run, explore, interact, and gain experience that may be of great value to future adoption candidates. Having a designated outdoor/exercise area (Figure 11.4) has been a highly appreciated addition to our facilities, not only for the dogs, but for the people as well. It has quickly fostered further transparency across our companies and the work we do every day. We have watched dogs completely change their "normal" cage behavior and become more bold, outgoing, playful, and adventurous. Dogs are let out all year round, and are able to experience different weather, smells, wind, sounds, sights, and textures such as falling leaves, snow, rain, etc. Watching dogs run, sniff, engage in play, and explore on open grass, apparently enjoying both the sun and the scents brought by the wind, gives a feeling of satisfaction that cannot be fully described with as much excitement as one feels when being a part of it.

Until the day comes where we have found alternatives to the use of laboratory dogs, we must continue to treat them with the very best care and compassion. We must be constantly thinking and striving to properly enrich them and excel at providing for their needs. Using dogs for our own benefits gives us a moral and legal obligation to ensure that the welfare of the dogs is given the highest priority in regard to both housing and use.

ACKNOWLEDGMENTS

Carolyn Allen would like to thank her staff and co-workers at AbbVie for their forward and progressive thinking and willingness (and support) to try new things when it comes to the realm of enrichment and welfare, as well as a heartfelt thank you to all of the laboratory animals who are ultimately benefiting humankind.

Jan Lund Ottesen would like to thank the animal caretakers working with dogs at Novo Nordisk for the commitment demonstrated in translating proposals from animal welfare experts into animal housing systems, socialization programs, and for building on their own experiences to look into new ways of working with dogs.

REFERENCES

1. Driscoll, C. A., Macdonald, D. W. & O'Brien, S. J. From wild animals to domestic pets, an evolutionary view of domestication. *Proceedings of the National Academy of Sciences of the United States of America* **106** Supplement 1, 9971–9978 (2009). doi: 10.1073/pnas.0901586106

2. Frantz, L. A. F. et al. Genomic and archaeological evidence suggests a dual origin of domestic dogs. *Science* **352**, 1228–1231 (2016). doi: 10.1126/science.aaf3161

3. Thalmann, O. et al. Complete mitochondrial genomes of ancient canids suggest a European origin of domestic dogs. *Science* **342**, 871–874 (2013). doi: 10.1126/science.1243650

4. United States Department of Agriculture: Annual Report Animal Usage by Fiscal Year. Available at: https://www.aphis.usda.gov/animal_welfare/downloads/reports/Annual-Report-Animal-Usage-by-FY2017.pdf (2018).

5. Commission Staff Working Document Accompanying Document to the Report from the Commission to the Council and European Parliament. Seventh Report on the Statistics on the Number of Animals used for Experimental and other Scientific Purposes in the Member States of the European Union. Available at: https://eur-lex.europa.eu/legal-content/EN/TXT/?uri=CELEX:52013SC0497 (2013).

6. Bradshaw, J. & Rooney, N. Chapter 8: Dog social behavior and communication. In Serpell, J. (Ed.), *The Domestic Dog: Its Evolution, Behavior and Interactions with People*, 2nd edition (Cambridge University Press, 2016), pp. 133–159.

7. Byosiere, S. E., Chouinard, P. A., Howell, T. J. & Bennett, P. C. What do dogs (*Canis familiaris*) see? A review of vision in dogs and implications for cognition research. *Psychonomic Bulletin & Review* **25**, 1798–1813 (2018). doi: 10.3758/s13423-017-1404-7

8. Kaminski, J., Schulz, L. & Tomasello, M. How dogs know when communication is intended for them. *Developmental Science* **15**, 222–232 (2012). doi: 10.1111/j.1467-7687.2011.01120.x

9. Hubrecht, R., Wickens, S. & Kirkwood, J. In Serpell, J. (Ed.) *The Domestic Dog: Its Evolution, Behavior and Interactions with People* (Cambridge University Press, 2016), pp. 271–299.

10. Jenkins, E. K., DeChant, M. T. & Perry, E. B. When the nose doesn't know: Canine olfactory function associated with health, management, and potential links to microbiota. *Frontiers in Veterinary Science* **5**, article 56 (2018). doi: 10.3389/fvets.2018.00056

11. Oesterhelweg, L. et al. Cadaver dogs - A study on detection of contaminated carpet squares. *Forensic Science International* **174**, 35–39 (2008). doi: 10.1016/j.forsciint.2007.02.031

12. Bijland, L. R., Bomers, M. K. & Smulders, Y. M. Smelling the diagnosis: A review on the use of scent in diagnosing disease. *Netherlands Journal of Medicine* **71**, 300–307 (2013).

13. Duranton, C. & Horowitz, A. Let me sniff! Nosework induces positive judgment bias in pet dogs. *Applied Animal Behaviour Science* **211**, 61–66 (2019). doi: 10.1016/j.applanim.2018.12.009

14. Range, F., Marshall-Pescini, S., Kratz, C. & Viranyi, Z. Wolves lead and dogs follow, but they both cooperate with humans. *Scientific Reports* **9**, article 3796 (2019). doi: 10.1038/s41598-019-40468-y

15. Hare, B. & Tomasello, M. Human-like social skills in dogs? *Trends in Cognitive Sciences* **9**, 439–444 (2005). doi: 10.1016/j.tics.2005.07.003

16. Kirchhofer, K. C., Zimmermann, F., Kaminski, J. & Tomasello, M. Dogs (*Canis familiaris*), but not chimpanzees (*Pan troglodytes*), understand imperative pointing. *PLoS ONE*, **7**(2), e30913 (2012). doi:10.1371/journal.pone.0030913

17. Miklosi, A. et al. A simple reason for a big difference: Wolves do not look back at humans, but dogs do. *Current Biology* **13**, 763–766 (2003). doi: 10.1016/s0960-9822(03)00263-x

18. Hare, B., Brown, M., Williamson, C. & Tomasello, M. The domestication of social cognition in dogs. *Science* **298**, 1634–1636 (2002).

19. Yong, M. H. & Ruffman, T. Emotional contagion: Dogs and humans show a similar physiological response to human infant crying. *Behavioural Processes* **108**, 155–165 (2014). doi: 10.1016/j.beproc.2014.10.006

20. Custance, D. & Mayer, J. Empathic-like responding by domestic dogs (*Canis familiaris*) to distress in humans: An exploratory study. *Animal Cognition* **15**, 851–859 (2012). doi: 10.1007/s10071-012-0510-1

21. *Guide for the Care and Use of Laboratory Animals (The Guide)* (8th edition) (Washington, DC: National Academies Press, 2011).

22. *Guidelines for Accommodation and Care of Animals (Appendix A) of the European Convention for the Protection of Vertebrate Animals Used For Experimental and Other Purposes* (Strasbourg, France: Council of Europe, 1986).

23. *Directive 2010/63/EU of the European Parliament and of the Council of 22 September 2010 on the Protection of Animals Used for Scientific Purposes* (European Union, 2010).

24. Clark, J. M. & Pomeroy, C. J. The laboratory dog. In Kirkwood, J. & Hubrecht, R. (Eds.), *The UFAW Handbook on the Care and Management of Laboratory and Other Research Animals* 8th edition (Hoboken, NY: Wiley-Blackwell, 2010).

25. Ottesen, J. L., Weber, A., Gurtler, H. & Mikkelsen, L. F. New housing conditions: Improving the welfare of experimental animals. *Atla-Alternatives to Laboratory Animals* **32** Supplement 1B, 397–404 (2004). doi: 10.1177/026119290403201s65.

26. Pacak, K. & Palkovits, M. Stressor specificity of central neuroendocrine responses: Implications for stress-related disorders. *Endocrine Reviews* **22**, 502–548 (2001). doi: 10.1210/er.22.4.502

27. Blanche, D., Terlouw, C. & Maloney, S. K. Physiology. In Appleby, M. C., Olsson, I. A. S. & Galindo, F. (Eds.), *Animal Welfare* 3rd edition (CABI, 2018).

28. *Committee on Recognition and Alleviation of Distress in Laboratory animals* (Washington, DC: The National Academies Press, 2008).

29. Beerda, B., Schilder, M. B. H., vanHooff, J. & deVries, H. W. Manifestations of chronic and acute stress in dogs. *Applied Animal Behaviour Science* **52**, 307–319 (1997). doi: 10.1016/s0168-1591(96)01131-8

30. Coleman, K. & Novak, M. A. Environmental enrichment in the 21st century. *Ilar Journal* **58**, 295–307 (2017). doi: 10.1093/ilar/ilx008

31. Gazzano, A., Mariti, C., Notari, L., Sighieri, C. & McBride, E. A. Effects of early gentling and early environment on emotional development of puppies. *Applied Animal Behaviour Science* **110**, 294–304 (2008). doi: 10.1016/j.applanim.2007.05.007

32. J.P., S. & J.L., F. *Genetics and the Social Behaviour of the Dog* (Chicago, IL: University of Chicago Press, 1965).

33. Jennings, M. & Howard, J. *LASA Guidance on the rehoming of laboratory dogs*, Available at: http://www.lasa.co.uk/wp-content/uploads/2018/05/LASA-Guidance-on-the-Rehoming-of-Laboratory-Dogs.pdf (2010).

34. Doring, D., Nick, O., Bauer, A., Kuchenhoff, H. & Erhard, M. H. How do rehomed laboratory beagles behave in everyday situations? Results from an observational test and a survey of new owners. *PLoS ONE* **12** (2017). doi: 10.1371/journal.pone.0181303.

12 The Non-Human Primate

Karolina Westlund and Lori Ann Burgess

CONTENTS

INTRODUCTION

In comparison to rodents and other smaller species, the proportion of non-human primates (NHP) used in animal-based research activities is minimal. For example, NHPs accounted for 0.2% of all animals reported in animal-based science activities in Canada in 2018, 9.6% of the non-rodent species reported to the United States Department of Agriculture (USDA) in the US in 2017, whereas they constituted on average 0.08% of animals used in the EU between 2015 and 2017.[1–3] Nevertheless, characteristics of the species require that particular conditions and care be in place to ensure the welfare of the animals. In this chapter, we discuss the two most common NHP families seen in the laboratory, macaques and callitrichids (marmosets and tamarins). For example, in 2017, cynomolgus (*Macaca fascicularis*), one of the macaque species, was the most commonly used NHP (88%) in European countries.[3]

BASIC BIOLOGY AND BEHAVIOR

The macaque family includes 23 species, only a few of those are taken to the laboratories. Macaques are ecologically diverse. In the wild, they are found in grassland, woodlands, and mountainous regions, as well as urban areas throughout most of Asia. They are diurnal, arboreal, and terrestrial. They are mostly herbivorous, feeding mainly on fruits, as well as seeds, leaves, flowers, and tree bark, although the crustaceans in their diet named the crab-eating macaque *(Macaca fascicularis)*. Rhesus macaques *(Macaca mulatta)* have specialized pouch-like cheeks, allowing them to temporarily hoard their food. While daily patterns vary, wild macaques forage and travel most intensively in the morning, and rest and allogroom throughout the afternoon.

Marmosets and tamarins are diurnal, arboreal, and are found in tropical rain forests in South America. Marmosets are omnivorous, although they primarily feed on tree saps and gums. They have long lower incisors which allow them to chew holes in trees to harvest the gum inside; some species are specialized gum feeders. This morphological adaptation, which is lacking in their tamarin cousins, allows them to inhabit more arid areas. In addition to tree saps and gums, wild marmosets spend some of their time foraging for insects. Similar to macaques, marmosets spend most of their time foraging and resting.

For most NHPs, vision is the dominant sensory modality. They are highly reactive to visual stimuli and make considerable efforts to gain visual information about their surroundings. In particular, they show a constant high level of attention to conspecifics – especially macaques. Macaques have full color vision (trichromacy), whereas among marmosets and tamarins, some females and all males exhibit dichromacy. They can all perceive ultrasounds (hearing range, macaques: 55 Hz up to 45 kHz; marmosets: 125 Hz up to 36 kHz), so monitoring the laboratory auditory environment outside the human hearing range is important.

Macaques, marmosets, and tamarins live in social groups and interact using a variety of facial expressions, vocalizations, body postures, and gestures. These differ widely between species, for instance, in rhesus and crab-eating (or long-tailed) macaques the bare-teeth display signals ritualized submission or fear toward dominant individuals, however, in Tonkean macaques (*Macaca tonkeana*) this facial expression signals affiliation. Marmosets and tamarins, in contrast, do not have very expressive faces but rather communicate through piloerection and the secretion of pheromones deposited in the territory through anogenital scent marking.

Macaques are highly intelligent and have demonstrated a variety of complex cognitive abilities, including the ability to make same-different judgments, understand simple rules, and monitor their own mental states.[4,5] Marmosets and tamarins are considered more "primitive," retaining some ancestral anatomical traits such as claws rather than nails, and being small-bodied they also have smaller brains and cortexes than macaques, both in absolute and relative terms.

SOCIAL DYNAMICS AND GROUP CONSTELLATIONS

There are substantial differences when it comes to social structure and behavior between the most common laboratory NHP species.

MACAQUES

In the wild, macaques live in social groups that can range from a few individuals to several hundred. Structure and size of groups vary across populations based on habitat and food availability, among others. In the laboratory environment, macaques are housed in multi-male multi-female groups if space allows, and single-male multi-female groups in smaller quarters; females should ideally stay in their natal groups whereas males are transferred to other groups at or around sexual maturity. Group housing is generally preferable over pair housing. If group housing is not a feasible option, it is typically easier to pair house two female individuals than two males together, although the success rate of the latter constellation can be increased by castration, by pairing an adolescent male with an adult, and through the process of non-contact familiarization (see below).[6] Another strategy that can be used under exceptional circumstances to avoid single housing is to pair compatible animals of the same genus but different species (e.g., *Macaca*).[7]

It is not always easy to find the right partner(s) for a particular primate. Everyone working with NHPs has stories or knows of one primate in their facility that has a more challenging personality and finding him/her that perfect partner is constantly on people's minds. The key is to not give up and label the animal as "non-pairable." It may take longer than one would like but having the mindset that "this animal has not found its right partner yet" leads to more constructive outcomes. If at first you do not succeed, do not give up, here is a good example why.

One facility had a cynomolgus (*Macaca fascicularis*) male, approximately five years old. He was a sweet monkey and was great with human interaction. However, he had to be singly housed upon arrival because no females were available, and it was difficult to find the right social setup for him. Several pairing attempts were made. Attempt #1: trio with two males (unsuccessful), attempt #2: pair with a male (unsuccessful), then attempt #3: another pairing with another male (also unsuccessful). After the third attempt, he was castrated to increase his pairing opportunities in the event that a female became available, and eventually one did. Attempt #4: paired with a female. It was obvious by his reaction when he was first introduced to her that he was very interested. Once they were paired, the bond was undeniable, within 30 minutes of being together, they were seen hugging and "kissing." In this case, it took three years to find the right, healthy match but it was worth all the work and time.

FIGURE 12.1 A marmoset family, as commonly seen in laboratory facilities, consisting of a breeding pair and several generations of twins.

Marmosets and Tamarins

In the wild, marmosets and tamarins live in family groups consisting mainly of one or two breeding females and a male and their offspring. They are characterized by a high degree of cooperative care of the young. Thus, marmosets and tamarins are ideally housed in monogamous pairs with offspring remaining in their natal groups even beyond sexual maturity (Figure 12.1).

Social Housing of Primates in Captivity

Generally speaking, for most species, when introducing unknown individuals to one another, it is a good idea to introduce them in neutral territory if feasible and place a temporary barrier between them that initially prevents physical contact. Evaluating their interactions and behavior towards each other will help to determine if proceeding is worthwhile. This type of *non-contact familiarization* allows animals to exchange important visual, olfactory, and auditory information and determine social rank. It has also been shown to greatly reduce injuries and be a good predictor of pairing success when used for pairing as reported for macaques.[6,8]

Single housing of primates is highly stressful and not conducive to good science, and thus should be avoided. The use of single housing must be justified, and is allowed only in exceptional circumstances by regulatory bodies.[9–11] If single housing is necessary, animals should be provided at least with olfactory, auditory, and visual contact with other animals in adjacent cages, additional human interaction and training as well as a generous enrichment program. However, consideration must be given to dominance and aggressive behaviors as some animals can impact the behavior of others, even if housed separately. For example, subordinate animals in adjacent cages may avoid locations close to their neighbors; therefore, it is important to assess the environment and ensure feeding and resting opportunities are not in areas avoided by these animals in such cases.

There are still concerns raised about social housing of primates, in particular for studies requiring implants and surgical procedures, or of infectious diseases. However, there are success stories about transitioning from single or pair to social housing. For example, one facility used to pair house long-tailed (*Macaca fascicularis*) and rhesus (*Macaca mulatta*) macaques involved in biosafety level 3 research (e.g., HIV-vaccine research). In the original setup five to seven cages were lined up along both long walls of a rectangular room, giving a total of 10–14 small cages/rooms. The decision was made to create a large, complex "extended" cage by opening up partitions between

FIGURE 12.2 Biosafety level 3 cages before and after being conjoined by tunnels across the room (images from two different rooms): a) Before adding the tunnels, animals were pair housed in two cage modules. b) After adding the tunnels, all separating walls were removed resulting in a seven-to-eight-fold increase in available space, tunnel space not included. c) The schematic drawing (which is not to scale) depicts space availability for one animal, in black, before (A) and after (B) the change.

cages and installing two tunnels to connect cages across the room, thus creating groups of 10–14 animals. These modifications brought changes on many fronts. Animal caretakers reported animals being happier, calmer, and more confident, and even approaching unknown people less hesitantly. They interacted more with humans and showed more greeting vocalizations than when they were pair housed. They also stopped dismantling water bottles, which used to be a big problem. The environment became less noisy and calmer. A more varied habitat could be offered, with bedding and foraging opportunities provided in the cages on one side, and different foraging opportunities, such as food puzzles as well as bathing options (baths offered about twice a week) available in those on the other side of the room. Also, there was less work involved in cleaning and the provisioning of enrichment materials. No systematic study was done, but anecdotally it seemed as if behavior became more diverse, fewer aggressive interactions were seen, and fewer injuries were reported. The latter may be explained by the fact that NHPs housed in the extended-cage version had several options for avoiding conflicts with other individuals with which they might not get along so well, and for spending time with friends (Figure 12.2).

THE ENVIRONMENT

The captive environment provided to NHPs should aim at stimulating them and improving their psychological well-being, while reducing fear, anxiety, and stress responses. This goal can be achieved by providing an environment that is sufficiently complex to enable the expression of many positive species-specific behaviors and provide a sense of security and control. The environment should also be adapted to the species' needs as space use differs between species, reflecting their respective ecological adaptations. Marmosets and tamarins are typically fully arboreal, whereas macaques spend a significant amount of time on the ground – yet will flee vertically when disturbed or frightened. For some species of macaque, the use of elevated structures is also related to the animal's dominance level, with higher-ranked animals occupying the highest elevations.[12] Thus, for all primates, providing usable three-dimensional space with adequate furnishings, including visual barriers and elevated platforms or areas which are big enough for a group of animals to huddle together and mutually groom each other, is extremely important. Another way of increasing stimulation and variety is to provide "windows" into the hallways of the animal facility or to the outside, if possible. Marmosets and tamarins also need at least one elevated sleeping box big enough for the whole group to fit in; typically, the whole family shares the same sleeping compartment in the wild. Macaques typically huddle together in small sub-groups overnight.

All primates should be provided with sufficient space to perform physical and social behaviors important to their welfare (e.g., grooming, resting, foraging, play, locomotion, etc.), while reducing

FIGURE 12.3 A) A biosafety level 1 volière for macaques. It includes objects that can be used as swings, perches, and elevated platforms, and branches for climbing, hanging, sitting, and moving in the vertical space. It also has windows to the outside. B) One of the benefits of a large space is that marmosets can leap from one platform to another.

the incidence of behaviors detrimental to their welfare, and thus meeting their needs. The ability to move about freely allows for more exercise and for the expression of a wide variety of natural behaviors (Figure 12.3). However, when providing larger spaces, it is essential to train the primates to voluntarily exit these areas on cue.

The environment should also give the animals a sense of security and control, and reduce fear and anxiety. Direct eye contact can be very challenging to primates – that is why they are often more comfortable looking at a person (or each other) through mirrors. Where hallway windows are provided, installing mirrors in the hallway can increase the viewing areas of NHPs, and even allowing them to see other animals in adjacent windows. Structures creating visual barriers can also be used as they provide some privacy and escape routes to avoid attacks and intimidation from other individuals. Another way of reducing fear is to provide NHPs with elevated areas (such as perches or platforms) where they can sit comfortably above the eye-level of humans – especially if they have not had much acclimatization or training (see Figure 12.4).

While a three-dimensional, structured environment is conducive to the expression of species-specific behaviors, it is important to ensure the objects and structures provided are safe and do not clutter the environment and restrict animal movements. Thus, a careful evaluation of the space must

FIGURE 12.4 Looking at a person through a mirror is less confrontational than direct eye contact.

be performed when introducing objects and structures. Similarly, an evaluation of the use of objects and structures, especially novel ones, should be performed.

FORAGING AND FOOD ENRICHMENT

In their natural environment NHPs will spend most of their waking hours foraging. Thus, in captivity, providing a variety of ways for NHPs to acquire their food will promote this important behavior. Emphasizing foraging and food enrichment will also contribute to improving animal welfare, especially if the space is minimal and limits the expression of other natural behavior. Feeding and food enrichment strategies should encourage NHPs to spend long periods searching for, gathering, and processing food. Strategies that can be used to increase foraging include providing food that is difficult to access (e.g., feeding through bars, feeding treats frozen in ice, and using puzzle feeders), and hiding small food items in the bedding substrate. Strategies and items selected should also take into account the feeding habits of the animals. For example, if they normally forage on the ground, food should be provided in a manner that functionally simulates such behavior.

It is known that providing valuable enrichment to NHPs can easily add up both in terms of money and manpower, depending on the size of the colony, partly because some of the devices sold, although convenient, can be costly. However, environmental and food enrichments contribute greatly to animal welfare, thus they should be an integral part of the management plan. Thus, finding creative, inexpensive and time-efficient ideas are primary factors in maintaining a successful enrichment program. If possible, recycling everyday materials is an easy way to add variety to the enrichment items provided and for promoting foraging in NHPs. A common practice that has proven to remain interesting over time, even after the novelty effect wears off, is the addition of foraging mix to the bedding. This practice has been shown to increase space use and behavioral variety, and to reduce aggression.[13,14] Because NHPs have enhanced tactile sensitivity[15] they benefit from objects and foraging opportunities that incorporate a variety of textures. As examples of effective, easy-to-implement and inexpensive foraging enrichment items that can be offered to primates, see Figure 12.5.

While using the creativity of the staff for developing enrichment devices and activities for the animals can contribute to improving the welfare of both humans and animals, one must ensure

FIGURE 12.5   A) A large space allows for larger and more creative enrichment, such as a bathtub. Although in the wild only long-tailed macaques have been documented to swim and Japanese macaques (*Macaca fucata*) to bathe in hot springs, rhesus macaques may also use, and enjoy, water pools to a certain extent. B) Recycled piece of plastic attached to the front of the cage and smeared with peanut butter and seeds on the outside can be used to occupy the monkeys for hours since they cannot see what is on the outside. It also allows animals to use their tactile abilities. C) Marmoset working for treats using PVC tubing filled with foraging mix.

that all the efforts are worth the work, and that the items meet their enrichment goal. This is one hard lesson learned at one facility. Every Friday at this facility, powder from the primate biscuits was collected to be used in creating elaborate "cookies" full of seeds, raisins, and other small food items. Preparing the cookies would take several hours, and then they would be put in the freezer. It was part of the Saturday duties for the animal care attendant to give out these frozen treats to the primates. The intention was that the primates would have to work hard at the frozen cookie to get all the little interesting bits out; it was thought that although time-consuming, the cookie preparation was worth it if the primates were spending time "foraging" on these cookies and enjoying digging out the hard-earned raisin or sunflower seed. Those cookies were integrated into the foraging enrichment program. Big mistake! No follow-up was done on how the primates used the cookies until many, many months later. When asked about it, the technicians mentioned that the primates just pop them in their mouths and ate them within a few seconds. Thus, the use of this "foraging" item was abandoned immediately since the time and effort put in the preparation was not eliciting hours of foraging behavior from the monkeys as expected. There were several lessons learned from this. Whenever implementing a new enrichment idea, it is essential to do a follow-up, or better yet, an evaluation on how the primates are using this new idea. Is it providing the primates the opportunity to express the natural behavior it was intended for? Is it providing any enrichment to their life? Is it contributing to improving animal welfare?

Provision of food enrichment, just like feed, must take into account dominance and competition among socially housed animals to ensure each animal's needs are met. Some facilities use positive reinforcement training when animals are housed in pens to keep the dominant individuals from harassing the subordinate ones during provision of food enrichment.

SOCIALIZATION AND TRAINING

Acclimatization and socialization protocols should be used to reduce fear of humans, and to prepare animals for research and other procedures where they may be in close contact with humans. Ideally, young animals should learn at an early age to associate human interaction with good, positive outcomes. It is presently unknown whether primates have an optimal socialization window comparable to that of cats and dogs. Nevertheless, acclimatization to humans should aim to prevent fear learning through the intentional use of latent inhibition, systematic desensitization, and counterconditioning (see Chapter 5).

Here are a few things to consider when establishing a positive relationship with primates:

- Avoid eye contact initially; move slowly and avoid making loud noises.
- Make a good first impression: avoid movements, postures, sounds, and objects that could be scary, especially the very first time you meet an animal; rather – bring food in order to establish yourself as a reliable predictor of good things through the process of classical conditioning. Even if some animals might be initially hesitant to take food in your presence, they still learn this association.
- Acclimatize animals gradually by using combined reinforcement (see detailed explanation below).
- Mimic friendly contact sounds or facial expressions (e.g., lip smacking, as seen in certain species of macaque) of the species in question.
- Once the animal is comfortably taking food out of the handler's hand, it is time to start systematic operant training.

Acclimatization need not be about relying on animals habituating to humans over time through a passive learning process. An intricate use of combined positive and negative reinforcement may be needed in some environments, particularly when primates are cage housed and cannot move away from the human handler. Instead, it is the human that moves away from the primate, contingently on the monkeys' behavior. One such procedure looks like this:[16]

- Place an assortment of tasty treats on surfaces near the cage front and back far enough away so that animals are not freezing from fear. Initially, the animals will likely jump away when approached. Avoid direct eye contact.
- As animals make any movement, back away even further until they have eaten the treats.
- Refill the treats and repeat the procedure, backing away contingently on animals' movement.
- Gradually shift the criterion so that backing away is contingent specifically on the animal's approach, not just any movement. Soon the animals may approach deliberately to make the trainer go away so they can get to the treats.
- As the animals gain confidence, gradually shape the distances shorter so that the trainer's initial, as well as final, positions are progressively closer to the cage.
- When the trainer is no longer backing away, offer the treats and let one hand remain hovering near the treats until the animals move forward – then slowly remove the hand. Repeat.
- Try offering the food out of the hand. This procedure may be successful quickly if the animals are interested in the food and moving. If carried out well, it is not particularly aversive to the animals since they are in control – by moving towards the trainer, they make the scary person go away, and get fabulous treats too!

Positive human interaction serves as a buffer that reduces stress reactions in primates, including aggressive responses to humans and abnormal behavior patterns, even before operant training has started.[16]

Training NHPs to participate in procedures (e.g., husbandry procedures such as shifting from one cage to another, research procedures such as injections and urine sampling and veterinary procedures such as attending to minor injuries) reduces stress experienced by the animals and decreases the risk of injury to both animals and personnel. While training increases the efficiency and ease of working with NHPs, it also increases cognitive stimulation and is a source of enrichment. Using the principles of operant conditioning, laboratory primates have been taught to participate in various medical and research procedures such as shifting enclosures on cue, stationing in specific locations, collaborating to perform specific responses together with cage mates, extending hands and feet for examination, taking oral medications, allowing ultrasound examinations, stethoscopes, thermometers, eye/ear/mouth examinations, blood sampling, injections, nose drops, and more. Also, they have successfully been trained for diverse operant tasks involving touch screens, levers, and other targets in cognitive/neuroscience research.

When working with marmosets it is imperative, as it is with any other species, to get voluntary cooperation from them in order to achieve the desired experimental results, while reducing the stress of the animals. Marmosets are a relatively "new" species in biomedical research, hence there is still a lot to be learned and standardized with them. Training procedures for marmosets have already been developed for simple tasks but we are finding that it may be possible for them to be trained using positive reinforcement training (PRT) for more complex tasks as well.[17] Research is showing that marmosets are capable of making simple cognitive associations and those can lead into potentially more complex ones all while using PRT to get them to achieve this.[18]

Similar to when working with any other species, it is important to determine what the preferred reward will be. An important consideration that needs to be made is the reward should be something that the marmoset can eat/drink quickly so that the momentum of training is not interrupted because he/she is running off with a piece of something to enjoy in the corner of the cage or enclosure. For example, being insectivores, they are very eager to work for a mealworm, however, they will grab this and run away to eat. This common response to mealworm reinforcers brings down the general rate of reinforcement, which may be impractical.

What has been found to work well is a reward in a liquid form that can be controlled by giving just a few small drops at a time. This is not too much that they get bored and are eager to try again and get a little more, so they are more likely to repeat the task being asked of them. Using a simple syringe with diluted syrup (to mimic the sweet sap from a tree they would normally forage for in the wild) does the trick. Placing a gavage needle on the end is helpful for ease of dispensing it directly to their mouth.

The duration of a training session can vary, but for increased success and maintaining their interest, it is best to keep it short, between 15 and 20 minutes. When marmosets are group housed the duration can go a little longer, as sometimes they will compete for training time. It is recommended to keep good records on each individual to help keep track of the stage of training, but also for recognizing the hierarchy within the group and which animal is allowed to train and for how long. For example, at one facility, a family unit of four marmosets was trained, two of them target trained very well while the other two were also good but were only allowed in the training box if the more dominant ones let them in. It is interesting to note that they did seem to learn from each other. Although the lower ranking marmosets did not get as much opportunity to perform target training, they were clearly observing their dominant counterparts and learned the task relatively quickly. Thus, their ability for social learning could be used to accelerate training for a task of a group of animals, or of shy individuals that are difficult to approach (Figure 12.6).

The latest trend in animal training aims to provide more control to animals in a training scenario. For example, the use of so called "Start Buttons" which provide greater control to the animals, allowing them to initiate trials within training sessions, is currently a hot topic in the animal training field. Perhaps by using such approaches, including giving choice in studies requiring many repetitions, would increase on-task performance. Indeed, self-paced unsupervised automated training looks very promising (for an example, see Calapai et al.[19]).

Due to the importance of training in reducing fear of humans and in habituating animals to procedures, an interesting venue of future research aiming at refining procedures and positively impacting human-NHP interactions might be to examine whether early socialization windows,

FIGURE 12.6 A) Training marmosets. B) Training tunnel in the indoor compartment of a biosafety level 1 housing. Visual barriers decrease social tension between animals. Animals are free to leave their respective training compartments at any time. C) Training macaques to present a leg for injection, accept an ear swab, open mouth for a saliva sample, and offer a hand for examination.

comparable to those found in cats and dogs, exist in primates. If so, those periods might be used to easily and effectively acclimatize young primates to humans, reducing the risk of later fear learning, and setting the stage for optimal human-animal relationships.

REFERENCES

1. United States Department of Agriculture. Annual Report Animal Usage by Fiscal Year. Available at: https://www.aphis.usda.gov/animal_welfare/downloads/reports/Annual-Report-Animal-Usage-by-FY2017.pdf (2018).
2. Canadian Council on Animal Care. 2018 Annual Animal Data Report (2019). Available at: https://www.ccac.ca/Documents/AUD/2018-Animal-Data-Report.pdf
3. European Commission. Report from the Commission to the European Parliament and the Council. 2019 report on the statistics on the use of animals for scientific purposes in the Member States of the European Union in 2015–2017. SWD (2020) 10 final.
4. Blanchard, T. C., Wolfe, L. S., Vlaev, I., Winston, J. S. & Hayden, B. Y. Biases in preferences for sequences of outcomes in monkeys. *Cognition* **130**(3), 289–299 (2014).
5. Couchman, J. J., et al. Beyond stimulus cues and reinforcement signals: A new approach to animal metacognition. *Journal of Comparative Psychology* **124**(4), 356–368 (2010).
6. Truelove, M. A., Martin, A. L., Perlman, J. E., Wood, J. S. & Bloomsmith, M. A. Pair housing of Macaques: A review of partner selection, introduction techniques, monitoring for compatibility, and methods for long-term maintenance of pairs. *American Journal of Primatology* **79**(1), e22485 (2017).
7. DiVincenti, L. Jr., Rehrig, A. & Wyatt, J. Interspecies pair housing of macaques in a research facility. *Laboratory Animals* **46**(2), 170–172 (2012).
8. Reinhardt, V. Pair-housing rather than single-housing for laboratory rhesus macaques. *Journal of Medical Primatology* **23**(8), 426–431 (1994).
9. Canadian Council on Animal Care (CCAC). *CCAC Guidelines: Nonhuman Primates* (Ottawa: Canadian Council on Animal Care, 2019).
10. National Research Council. *Guide for the Care and Use of Laboratory Animals* (Washington, DC: National Academy Press, 2011).
11. European Commission. Directive 2010/63/EU on the protection of animals used for scientific purposes. *Official Journal of the European Union* **276**, 54–71 (2010).
12. Reinhardt, V. Space utilization by captive rhesus macaques. *Animal Technology* **43**(1), 11–17 (1992).
13. Chamove, A. S., Anderson, J. R., Morgan-Jones, S. C. & Jones, S. P. Deep woodchip litter: Hygiene, feeding, and behavioral enhancement in eight primate species. *International Journal for the Study of Animal Problems* **3**(4), 308–318 (1982).
14. Doane, C. J., Andrews, K., Schaefer, L. J., Morelli, N., McAllister, S. & Coleman, K. Dry bedding provides cost-effective enrichment for group-housed rhesus macaques (*Macaca mulatta*). *Journal of the American Association for Laboratory Animal Science* **52**(3), 247–252 (2013).
15. Hoffmann, J. N., Montag, A. G. & Dominy, N. J. Meissner corpuscles and somatosensory acuity: The prehensile appendages of primates and elephants. *The Anatomical Record, Part A: Discoveries in Molecular, Cellular, and Evolutionary Biology* **281**(1), 1138–1147 (2004).
16. Westlund, K. Training laboratory primates–benefits and techniques. *Primate Biology* **2**(1), 119–132 (2015).
17. McKinley, J., Buchanan-Smith, H. M., Bassett, L. & Morris, K. Training common marmosets (*Callithrix jacchus*) to cooperate during routine laboratory procedures: Ease of training and time investment. *Journal of Applied Animal Welfare Science* **6**, 209–220 (2003). doi: 10.1207/S15327604JAWS0603_06
18. Miller, C. T., Freiwald, W. A., Leopold, D. A., Mitchell, J. F., Silva, A. C. & Wang, X. Marmosets: A neuroscientific model of human social behavior. *Neuron* **90**(2), 219–233 (2016). doi: 10.1016/j.neuron.2016.03.018
19. Calapai, A. et al. A cage-based training, cognitive testing and enrichment system optimized for rhesus macaques in neuroscience research. *Behavior Research Methods* **49**(1), 35–45 (2017).

13 The Pig

Mette S. Herskin, Cathrine Juel Bundgaard, Jan Lund Ottesen, Dorte Bratbo Sørensen, and Jeremy N. Marchant-Forde

CONTENTS

THE PIG

Due to the large anatomical and physiological homologies with humans, the use of pigs for research purposes is increasing.[1] Even though pigs are relatively new as laboratory animals, the human-pig interface goes back a long time and is dominated by the use of pigs for human consumption. Across languages, the traditional understanding of pigs as dirty and not bright is evident from popular sayings, such as: "dirty as a pig," "I sweat like a pig!" "you are dumb as a pig," or "you eat like a pig." This chapter reviews the biology of the pig, including natural behavior, cognitive as well as emotional abilities, and challenges the validity of the popular sayings, in a discussion of the possibilities to house, manage, handle, and train pigs in laboratories. Throughout, the chapter presents possibilities to improve the welfare of pigs, based on a combination of knowledge from studies of farmed as well as laboratory pigs and practical experience working with and training pigs kept for research purposes. One major aim of the chapter is to remind readers that pigs are not large mice, or dogs with less hair. Pigs are not humans either. Pigs are pigs – irrespective of the breed. Constantly remembering this can be one big step in the direction of improved welfare of this species.

CHARACTERIZATION: MORPHOLOGY, DOMESTICATION, AND BREEDS

Taxonomically, pigs belong to the order of even-toed ungulate mammals and the genus *Sus*. All domestic and miniature pigs share the wild boar as an ancestor. Today, more than a hundred pig breeds of rather different morphology are available, mainly differing in size at sexual maturity, growth potential, and reproductive biology. As part of domestication, modern pigs have been bred for their growth potential. This means, that a mature sow of a meat production breed is much larger than her ancestor (300 kg versus less than half for a wild boar sow) and that a growth rate of around 1 kg/day is not unusual. Among the many breeds, commercial farm breeds (such as Yorkshire, Duroc, or Landrace), Yucatan miniature pigs, and Göttingen minipigs are most often used for research purposes. A detailed list of different breeds and their characteristics can be found elsewhere.[2]

The Biology and Natural Behavior of the Pig

Even though laboratory pigs most often are kept in conditions differing substantially from the natural conditions of pigs, knowledge about the natural behavior of pigs is important in order to understand and improve the welfare of pigs kept for research purposes. As mentioned, the breeds typically used in laboratories may not look very similar to the wild boar, but studies of the behavior of feral pigs or domestic pigs under semi-natural conditions[3] have shown that their behavior shares many similarities. Below, selected parts of natural pig behavior, considered relevant for laboratory pigs, are presented. A thorough review of the natural behavior of pigs can be found in D'Eath and Turner.[4]

Pigs can thrive in a variety of habitats. Pigs are not strongly territorial, they merely defend resources within their home range. Under laboratory conditions, it is thus important, in terms of animal welfare, to pay attention to limitations in the access to resources – be it restricted feeding, restricted number of eating places, or for example, restricted access to manipulable materials. Restricted access to resources may lead to aggression and fighting and thus challenge the welfare of the pigs.

Pigs are social animals, the group size of which is flexible, depending on the habitat type and resource availability. However, except for mature boars, pigs are not solitary and social isolation is a severe stressor, especially for juveniles.[5,6] In nature, the typical core group consists of mature sows (often genetically related) and their piglets from the most recent breeding season. Piglets are weaned gradually at three to four months of age. When pigs are produced commercially for meat, weaning is abrupt and takes place much sooner than under natural conditions (typically when the piglets are three to five weeks old). Hence, it is important to bear in mind that even though pigs in the laboratory purchased from commercial farms may weigh 20 kg or more, they are still juveniles, and in nature they would not have yet been weaned, and may thus have other needs than adult pigs.

A sow is pregnant for 114–118 days and domestic sows have been selected to be able to breed year-round. The litter size of the wild boar varies from four to six, whereas sows of meat-producing breeds may give birth to litters of up to more than 20 piglets, depending largely on the breed. In natural conditions, sows will leave their group one to three days before parturition and seek out a suitable birth site. One major characteristic of the maternal behavior of pigs – in all environments – is their strong motivation to build a nest before giving birth. The documented strength of this motivation[7,8] has led to international legislation, e.g., in the EU, which includes the provision of nesting materials to sows/gilts in the days before giving birth (farrowing).

The post-partum maternal behavior of sows is unique among mammals. Piglets are born precocial and can walk and seek the udder without any maternal assistance. Sows do not lick their offspring – the main maternal behavior of a sow at birth is to provide an accessible and warm udder to the piglets. In the first days after birth, a sow and her piglets will stay in or near the nest, providing thermal protection for the piglets. Later, the sow brings the piglets back to the group, into which they will gradually integrate without any aggression. After weaning, female offspring will either stay in their maternal group or form a new group, and males disperse over greater distances. Sexually mature boars form smaller groups or live solitarily except for the mating season.

Within a group of pigs, a stable hierarchy exists, and high social status is conferred by maturity, a large body mass, and physical strength. Aggressive interactions over resources such as food can occur between group members but is often limited. Serious fighting can, however, break out when unfamiliar pigs meet, especially in a confined space. After mixing with unfamiliar individuals, pigs of all ages will seek to establish a new hierarchy, which can lead to severe and prolonged fighting, especially if the pigs are of similar age and weight (as it is often practiced in laboratories and pig barns) and if mutual avoidance is not possible. Hence, in order to improve the welfare of pigs used for research purposes, mixing of unfamiliar individuals should be limited as much as possible. In a group of pigs, cohesion is enhanced by social facilitation, which means that the pigs will synchronize their activities. Thus, the welfare of pigs is improved if all individuals in a pen are able, for example, to eat, sleep, or manipulate materials together.[9]

The diurnal rhythm of domestic pigs kept in semi-natural conditions is characterized by relatively long periods of rest/sleep and long periods of activity. During the active periods, the pigs spend more than half of their time in foraging-related activities including rooting, grazing, and exploring substrates with their snout. Pigs are omnivorous, opportunistic, and highly explorative. This means, that even though laboratory pigs may be fed a nutritionally rich and optimized diet, which they can eat in a rather short time, they will still be highly motivated to perform exploratory behavior for several hours each day. Below, knowledge about enrichment of pig housing is reviewed.

The pig snout is sometimes called their "most important sensory organ." The snout is the elongated nose, very sensitive to tactile, gustatory, and olfactory stimulation and highly innervated, ending in a cartilage disc with central nostrils. However sensitive, the snout of pigs is very strong and flexible, and used a lot to manipulate the surroundings. In addition to the snout, one other special characteristic of pigs, deviating from most other mammals, is the lack of sweat glands. This means, that pigs are vulnerable to heat stress and will need to use behavioral means of temperature regulation. Hence, contrary to popular sayings about sweating and dirty pigs, pigs cannot sweat like humans, and being able to wallow in mud is one prioritized way of cooling down for pigs. If wallowing is not possible, alternative means to cool down may be to lie laterally and without physical contact with other pigs. The thermoneutral zone of pigs is age-dependent – the younger the pigs, the higher temperature they need (Table 13.1).

COGNITIVE AND EMOTIONAL ABILITIES

Taking the long-term shared history of humans and pigs into consideration, the current knowledge about cognitive and emotional capacities of pigs is strikingly low.[12] However, what is known suggests that pigs are cognitively complex and share many traits with animals considered intelligent, such as marine mammals. Recently, observations of unprompted instrumental object manipulation, corresponding to tool use, was reported in Visayan warty pigs (*Sus cebifrons*) during nestbuilding.[13]

Among the non-social cognitive abilities of pigs are object discrimination, which has been demonstrated in situations requiring considerable memory.[14] Pigs, as foraging omnivores, use spatial information for discrimination and learn relatively easily to navigate and remember mazes and other spatial settings.[15] Moreover, pigs have been shown to be able to use a mirror image to locate hidden food.[16] Equipped with well-developed abilities for remembering where food patches are, pigs augment these by using the behavior of group-mates as further sources of information about the location of food, and even seem to be able to use it to their own advantage, e.g., by practicing

TABLE 13.1
The Thermoneutral Zone of Pigs (the Range of Environmental Temperatures within Which Metabolic Rate and Heat Production Are (Fairly) Minimal, Constant, and Independent of the Ambient Temperature) Depends on the Breed and Age of Pigs. The Table Shows Selected Examples

Farm Pig	Thermoneutral Zone	Göttingen Minipig	Thermoneutral Zone
Adult.	10–21°C	Adult.	15–25°C
Grow/finish pig.	10–21°C	Adolescent (14–16 weeks).	15–25°C
Nursery pig (20–35 kg).	15–21°C	Piglet (6–8 weeks).	20–30°C
Nursery pig (10–20 kg).	21–27°C		
Weaner (3–4 weeks).	24–30°C		
Newborn piglet.	32–38°C		

Sources[10,11].

intentional deception.[17,18] Hence, pigs seem to be as socially complex as other highly intelligent mammals.[19] For instance, pigs discriminate among conspecifics and prefer familiar individuals over strangers[20] and sows are able to discriminate their offspring from other litters by voice.[21] Thus, pigs appear to have strong abilities to flexibly discriminate among conspecifics using various cues and under a variety of circumstances. In addition, pigs can discriminate familiar and unfamiliar humans,[22] discriminate different attention states in humans[23] and use pointing by humans to lead them to a food reward.[24]

Knowledge about cognitive abilities does not only come from carefully designed ethological experiments. Recently, incidental observations of feeding behavior of wild boars in a zoo, bringing apple halves soiled with sand to a water source and washing them before eating,[25] suggest that pigs can discriminate between soiled and unsoiled foods and that they are able to delay gratification for long enough to transport and wash the items. So far, such abilities have only been described in primates and some birds.

Nowadays, not only cognitive abilities, but also the emotional states of pigs receive scientific attention and method development.[26] Recently, evidence for emotional contagion in pigs has been put forward, suggesting that not only can pigs connect with the emotions of other pigs; they can do so with pigs who are responding emotionally in anticipation of future events.[27] Thus, when group-housed pigs are repeatedly submitted to certain types of stressors (e.g., catching, restraint, or blood sampling), they experience contagious fear when they witness the stress of others, and the contagion is potentiated by previous exposure to the same stressors.[28]

One way to assess the emotional state of animals is by use of tests of cognitive bias.[29] By use of this methodology, it has been shown that environmental enrichment (a combination of space, straw, and objects) induced "optimistic" cognitive bias in pigs, indicative of a more positive emotional state.[30] Cognitive bias testing has also been used in studies of animal pain, but so far not in pigs, and it is striking to see how little the pig is represented in veterinary textbooks on pain management. The majority of knowledge about porcine pain comes from studies of so-called management routines, practiced in commercial pig production, such as tail docking or castration. A recent study documented a rather low reporting of pain relief in pig studies involving experimental surgery.[31] There are, however, studies of pain in pigs used as surgical models for humans,[32] and studies of pain in pigs used as models for naturally occurring pain states, such as lameness[33]. Recently, work has been done to develop and validate piglet grimace scales,[34,35] but they are not yet as operational as the comparable rodent scales. A review of pain indicators, including research models, procedures common to pig production and naturally occurring tissue damage can be found in Herskin and Di Giminiani.[36] Due to the unambiguously negative relation between pain and animal welfare, we strongly believe that pigs should be given the benefit of the doubt in cases where pain is suggested to be involved.

THE ENVIRONMENT

Housing for the laboratory pig often meets the concept of traditional laboratory animal housing, i.e., housing that is sterile, based on single housing, easy to manage, environmentally controlled, and built for the specific needs of the caretaker and for the procedures which the pig will undergo. There is also the widely held belief that for laboratory animal studies to be reproducible there should be environmental standardization and that this is best achieved by having identical housing conditions. However, the same environments may actually be detrimental to experimental reproducibility[37] and systematic variation of environmental conditions such as enrichment items, enclosure size, sound, and lighting, may be beneficial to produce robust and valid results.[38]

As can be seen in the previous sections, the pig is a complex animal, and his/her welfare is best served by an environment that matches that complexity. The pig is social and has a very keen sense of smell. The pig spends a lot of time resting and foraging and the main thermoregulation mechanism is behavioral. Adding these all together, standard laboratory housing is likely to result in welfare issues, unless structures or processes that better meet the pig's needs are incorporated.

In principle, the easiest way to achieve complexity is by including outdoor access. Outdoor access can be in the form of an "activity area" – perhaps with access to a pool – or as small enclosures (Figure 13.1, Figure 13.2). Of course, hygiene and ease of management are valid and important factors for pig welfare, but environmental complexity does not have to mean that welfare is jeopardized, and overall, can mean the opposite. Elements of complexity may be grouped together in terms of comfort, enrichment, and the social environment, and will be discussed as such below.

AN ENVIRONMENT DESIGNED FOR COMFORT

To enhance animal welfare, laboratory animals should be provided with bedding materials or resting structures adapted to the species and a solid, comfortable, clean, and dry resting area shall be provided. In the EU directive 2010/63/EU, specific regulation has been laid down for this. Moreover,

FIGURE 13.1   Outdoor exercise and exploration area with a pool. The pool can also be used for wallowing.

FIGURE 13.2   Outdoor rooting area consisting of a shallow pool filled with bark and wood chips. Food treats can be mixed into the chips to promote foraging behavior. This area can be used when pigs arrive to allow them to settle down before they are allowed access to the indoor pens.

temperature and relative humidity in the holding rooms must be adapted to the pigs housed. The chosen temperature should lie within the optimal part of the thermoneutral zone (the comfort zone), which will vary based on age, weight, and breed of pig – i.e., whether it is a meat production breed or a minipig (Table 13.1). Clearly, housing different aged pigs within the same room will be problematic, unless the pigs themselves are able to adjust their thermal environment. It is important to keep in mind that adult pigs, and especially lactating females, are prone to heat stress, and their upper limit is quite low.[39,40] Thus, even if the room temperature is within the thermoneutral zone, incorporating elements that provide pigs with choices will be beneficial. This could be achieved by either providing zones with differing temperatures – pigs will choose temperatures that are more comfortable[41] – or by maintaining temperatures and offering choice in floor substrate. For juvenile pigs in particular, part of the pen can be equipped with supplementary heat, in the form of a heat lamp or a heated mat. This gives a localized hot zone, which should be big enough for all the pigs to fit underneath if they want to, but also the ability to get away and choose a cooler area. Having a mixture of solid and slatted or perforated flooring can also give pigs some thermal choices, with slatted floors generally a few degrees cooler than solid floors, due to airflow from underneath. Solid floors also give the option of adding a bedding substrate, which may not only confer advantages in terms of thermal comfort, but can also offer physical comfort and be a source of environmental enrichment (see below).

Hard concrete, metal, or plastic flooring are known risk factors for physical injuries such as decubital shoulder ulcers,[42] claw lesions,[43] and forelimb skin abrasions in pigs.[44] Adding a bedding substrate such as straw, shavings, or sawdust must thus be prioritized as it can decrease the prevalence of the above-mentioned clinical conditions. Bedding substrate may also reduce the prevalence of other conditions such as leg and hoof injuries, damaging tail-biting behavior, and gastro-intestinal disorders, including stomach ulcers.[45] Moreover, bedding will improve the physical and thermal comfort of the pig. If the pig facility is a bio-secure environment, any organic bedding may need to be sterilized prior to introducing it into the facility. However, sterile bedding may be hard to locate. An alternative could be the use of rubber or plastic-coated foam matting to improve comfort and improve various measures of welfare compared to concrete floors with no bedding.

An Environment with Enrichment

Keeping pigs in barren environments can lead to behavioral problems, lack of stimulation, and poor welfare, but these issues can be improved by the use of environmental enrichment. Environmental enrichment can be defined broadly as "an improvement in the biological functioning of captive animals resulting from modifications to their environment."[46] The key part of this definition is the improvement in biological functioning. The addition of any complexity into a barren environment does not necessarily improve it, and complexity must be reviewed in terms of biological relevance to the animals and its ability to improve their welfare.

So, for the pig, what constitutes an enrichment? What is biologically relevant to the pig, beneficial to the pig's welfare and rarely, if ever, associated with welfare problems? For an enrichment to be relevant, it needs to have properties that allow a pig to express key elements of her/his behavioral repertoire.[47] Much of the focus with pigs has been on items that address their foraging and exploratory nature. Hence, manipulable items or the possibility to perform foraging or cognitive activities should be provided to ensure the welfare of pigs, irrespective of the environment where the pigs are kept. The main characteristics of enrichments are that they are: 1) edible (pigs can eat at least parts of it), 2) chewable (pigs can bite and chew them), 3) investigable (pigs can tear them apart), and 4) manipulable (pigs can change their location, appearance, or structure).[48,49] Additionally, enrichment should be provided so that they are of sustainable interest, given in sufficient quantity and be clean and hygienic. Many of the "enrichment" items sold by laboratory animal equipment retailers often address only one or two of the key characteristics in that they are chewable and/or investigable.

They will get some interest from pigs when added to a barren environment but the motivation to interact will decrease over time as the novelty wears off.[50] However, it is possible to design enrichment items that stimulate exploration shown as rooting, chewing, eating, and moreover can be baited with different foods to further facilitate these behaviors. One example is a snout-operated pellet dispenser, which can be made from a section of PVC pipe or similar, adding cap ends and drilling holes throughout the PVC. The pipe is then filled with treats and can be left on the floor or mounted on the wall in a way to enable it to rotate (see photo and detailed description in Marchant-Forde and Herskin[51]). Pigs will have to root and turn the pipe to allow the treats to fall out. Another option is the use of separate pens to become a designated "activity room" through which pigs can be rotated for specific periods of time.[52]

When laboratory pigs are purchased from private farms, the pigs may have been tail-docked as juveniles. Routine use of this procedure, done to avoid later outbreaks of tail-biting behavior, is illegal in the EU.[53] Importantly, raising meat production breed laboratory pigs in spacious pens with sufficient bedding, rooting material, and conscientious management may eliminate the need for tail docking as tail biting ceases to appear (personal communication, Jan Lund Ottesen, 2019).

A Social Environment

Within the laboratory setting, pigs may be relatively long-term residents and during this time undergo repeated disruption of stable groups and periods of single housing (with or without auditory, visual, and olfactory contact to other pigs). The welfare consequences of group disruption (when one or more individual pigs leave a stable group) have not been studied in depth. However, both mixing of unfamiliar pigs, introduction of new individuals into established groups, and different levels of social isolation are social stressors with a relatively large influence on the behavior and physiology of pigs, thereby impacting their welfare[54] and, potentially, also the outcome of scientific studies.

Whenever pigs are mixed, aggression is inevitable as they form their social hierarchy. Usually, the level of aggression, and especially the consequences in terms of injuries, are worse, the older the pigs when mixed. Hence, it is recommended not to mix adult pigs. There are, however, methods to reduce aggression at mixing.[55] For example, the use of pen designs incorporating extra space and, if possible, getaway areas/barriers to retreat behind. Additionally, the use of management techniques such as provision of food ad libitum, mixing in the presence of a super-dominant individual, pre-exposing pigs to each other by penning in adjacent pens, and allowing litters to mix while young to build social skills. Once the hierarchy is established, overt aggression should be seen only occasionally.

Recommendations for laboratory pigs are, wherever possible, to house pigs in groups or pairs and the importance of social support cannot be understated.[56] Although social housing is preferred, there will be studies and procedures in the laboratory setting, that require pigs to be housed individually. For example, pigs that have undergone implantation of catheters or access ports will need to be housed singly to prevent pen mates from damaging both the implant and the host pig. Such periods involving single housing should be kept as short as possible. So far, few studies have sought to optimize housing of laboratory pigs in situations, where single housing is required[5] and best practice probably depends on the age and breed of the pigs involved. In a study of newly weaned pigs of a meat producing breed, kept individually with/without surgical catheterization, and with/without contact to conspecifics through wire mesh, Herskin and Hedemann[57] found that allowing the contact through wire mesh resulted in increased activity and increased play. These findings indicate that the provision of limited social contact may help reduce the negative effects of individual housing. Results of a choice experiment[58] suggested that access to a mirror may be used by pigs for social support during periods of perceived threat (when a human is in the room). In a later study, the authors found indications of visual contact as being more of a frustration in situations where physical contact was not possible.[59] Hence, even though further studies are still needed in order to

FIGURE 13.3 Pigs housed in flexible pens. In this picture, three pens are combined to provide one larger pen for group-housed pigs. The pens are modified by removing/installing the middle part of the separation walls. In addition to wire mesh allowing pigs to see each other, when single housed, the solid walls have holes (A) which the pigs use for snout-to-snout contact.

provide a fully science-based recommendation for single housing of laboratory pigs, we do recommend that at least physical contact through wire mesh or similar should be provided when possible (Figure 13.3).

BEHAVIORAL MANAGEMENT (HANDLING, HABITUATION, AND TRAINING)

Most pigs used for research are meat-producing breeds purchased from farmers or special-purpose breeds like the Göttingen minipig. Purpose-bred pigs have often been socialized from birth to accustom them to human presence, but all pigs need socialization and habituation when they arrive at an experimental animal facility. Below, we discuss different options for these procedures based on the personal experience of the authors and their coworkers.

Stress associated with transportation and a new environment has widespread effects on the physiology of pigs.[60,61] These changes can confound research if studies are initiated before homeostasis is restored and physiological measures are returned to baseline. Therefore, some period of acclimatization is generally suggested. The standard is usually one week but depending on what needs to be obtained in the acclimatization period, the required period may be longer.[62] The acclimatization period can be used efficiently for socialization, habituation, and training the pigs for the daily husbandry routines and the procedures involved in the study in which they are to be enrolled (see Chapters 5 and 6 for details on animal learning and training). Using the extra time can markedly enhance pig welfare, ease handling, and save time during the study. Moreover, it may minimize inter-pig and group variability in the studies. Further research is needed, though, to establish the effect of positive reinforcement training on variability and data quality in pig models.

It is normal procedure to have thorough protocols for animal research. We recommend developing equally thorough protocols or programs for the socialization, habituation, and training of the pigs including step-by-step training protocols, tutorials, and a log system. The programs should be designed to minimize stress in the pigs throughout their stay in the facility. Pig socialization and training programs provided by skilled staff, will make the procedures and handling less stressful, and, all other things being equal, less stressed pigs should result in better research models.

Furthermore, it will improve the working environment for the staff since the cooperation between animal and handler is often considered more satisfying than using restraint and force.

Socialization, habituation, and training can be initiated as soon as the pigs arrive. The first day the pigs will be stressed from the transportation and the novelty of surroundings and perhaps new groupmates. It is important that the arrival is as calm as possible.

If the pigs are fighting on arrival, chemical intervention such as spraying with a peppermint solution or adding novelty such as access to a rooting area with wood chips (Figure 13.2) may reduce aggression. However, these effects are often short-term, and it must be ensured that the housing pens are of sufficient size and enriched to avoid any initial aggression to continue.[55] Furthermore, it is helpful if the pens are flexible so more space can be provided when receiving new pigs to the facility, and thus allow for group housing during the acclimatization period (Figure 13.3). If or when the pigs need to be single housed (e.g., if they get catheters inserted) the pens can be modified to accommodate this and hence the use of flexible pens reduces the need for relocation of the pigs when changing from group to single housing.

Despite the omnivorous nature of pigs, they may show some degree of neophobia, and thus a change in diet may take some time, especially if they have been reared exclusively on lab or commercial pig diets. Therefore, new food items that later will be used when training the pigs, should be introduced as soon as possible. When familiar with the food items, pigs will work for access to many different types of food (such as apple, banana, raisins, grapes, dog treats, uncooked pasta, commercial pig food, and chocolate). If a pig is highly reluctant to try new, chewable food, gently spraying apple juice into her mouth may quickly prompt her to eat small pieces of apple. From a training perspective, it can be an advantage to use fluids such as juice or yoghurt as reinforcers (see also Chapters 5 and 6). The fluid is given to the pig using a bottle with a dispenser (Figure 13.4), which makes it easy to reinforce the pig. Throwing food reinforcers on the floor may result in the pig engaging in rooting behavior that may slow down the training. Hand feeding the reinforcers may result in the trainer being bitten if the pig is very eager.

The amount of time needed to socialize and train pigs may differ between breeds and individuals. Socialization can be started the day after arrival. Since pigs are normally kept in rather limited space in the pens, it is a good idea to use the corridors of the facilities as exercise areas (Figure 13.5). This may also help the caretakers when cleaning pens and it is an excellent possibility for habituating and socializing with the pigs.

FIGURE 13.4 Positive reinforcement training using a target-on-a-stick. The target (A) is presented to the pig. If the pig touches it without biting it, the trainer will mark this behavioral response by clicking using a clicker (B). The click will be immediately followed by a reinforcer; in this case some yoghurt from a plastic bottle (C).

FIGURE 13.5 During pen cleaning, the pigs can be allowed to roam the corridor to play and explore. Whether the pigs are allowed to go in groups or one at the time, depends entirely on group dynamics. Normally, single-housed animals are corridor walked alone. The corridor playtime is also a chance for the caretaker to interact with and socialize the animals.

On the second day, time should be spent with the pigs, feeding them treats. If the pigs are very large, fearful, or aggressive, feeding can be done from outside the pen. The barrier will keep the caretaker safe and the pig may be less fearful if the human is not inside the pen. If possible, the caretaker should enter the pen several times during the day. Pigs are naturally exploratory and sitting or squatting will often make even very timid individuals approach the caretaker. When the pig approaches, the caretaker can try to touch him/her gently, but the pig should be allowed to withdraw and should never be chased. We recommend talking gently to the pigs and feed treats on the floor as a start. Hand feeding the pigs may be risky as some individuals may be eager and bite the hand when taking the food. Later, when the pigs have learned that the caretaker brings treats, some pigs may start to be rather intrusive. It is thus important never to reinforce any behavior such as biting or pushing. Additionally, the pig can be trained to back away from the gate or to follow a target, which will allow the trainer to guide the pig away from the gate so that the trainer can enter the pen (Figure 13.4).

Pigs should be trained using positive reinforcement (see Chapters 5 and 6). The use of positive reinforcement training will develop a positive human-animal relationship with safe and positive interactions, where pigs voluntarily cooperate. In that way restraint can be reduced or eliminated. Prior to training, the trainer must decide on which reinforcers (treats used for specific training purposes) to use and make sure that the pig eats the reinforcers at a steady rate. If a conditioned reinforcer (see Chapter 6 for details) such as a clicker or a whistle is to be used, it should be conditioned when the pigs are confident with the trainer and eat the offered food without hesitation. The association between the conditioned reinforcer and the primary reinforcer (the food) is very quickly established in pigs. Often it is possible to condition the clicker/whistle and start target training within the first session.

We recommend that pigs are trained twice a day on weekdays with a pause on weekends. Group- or pair-housed pigs should be separated during training sessions. Alternatively, two or more trainers should be working with the pigs; one trainer working with the pig being trained and the other trainer(s) feeding the other pig(s) making them stay away from the pig being trained. Each training session should be approximately two to three minutes per pig and time should be allocated for the training, so it becomes a daily task ranging equally with other tasks such as cleaning and feeding.

Moreover, if the training is to be successful, a thorough, continuous, and qualified education of the trainers is needed.

Pigs can learn a large variety of behaviors. Following a target can be used to facilitate weighing (the pig will enter the scale) or local transportation (the pig will enter a transportation crate) or for gating, i.e., going from one place to another. Most pigs will learn to follow a target in two to three days. Pigs may also be trained to stand still and be calm while ointments are applied, injections given, vests or loggers of different types attached to the pig, or while a blood sample is drawn from an implanted catheter. These behaviors may take up to a couple of weeks to train depending on the animals and the skills of the trainers. Pigs in the authors' facility have been trained to stand immobile for up to ten seconds multiple times in a special transportation crate while x-rays are being taken.

It is not necessary to have the same trainer all the time, but experience from other species indicates that the animal will learn the task faster if it is only one person acting as a trainer in the learning phase. Hence, the rule-of-thumb is "one new behavior – one trainer only." When complex behaviors are trained, this rule should always apply. In any case, it is important to have a thorough protocol for the training, and to have a log, to pass on information about the training from one trainer to another (please refer to Chapter 6).

Pigs that are routinely trained and socialized are less fearful of humans and they are expectant and calm when people enter the room. Well-socialized pigs often like to be scratched on the body with a brush and scratching may even be used as a reinforcer. Many pigs will lie down on their side when the trainer or caretaker gently rubs their belly. Recently, it has been shown that physical human-animal interactions may be important by documenting that belly rubbing (abdominal stroking) was able to induce a positive welfare state interpreted via electroencephalogram (EEG) studies.[63]

Positive reinforcement training provides the pigs with control in potentially aversive situations, as the trained pig can predict the events and choose to participate. Hence, the training may eliminate the need for restraint and other forceful handling techniques normally described in the laboratory animal science literature. It therefore markedly reduces the amount of negative interactions experienced by both animals and humans.

WHERE TO GO NEXT?

It is important to keep in mind that "the laboratory pig" covers multiple areas of research – from classical animal models for humans to porcine models for porcine conditions. Notwithstanding the purpose for which the pigs are kept, pigs are intelligent, social animals with a complex behavioral repertoire reminiscent of their ancestor, the wild boar. In order to maximize pig welfare, anyone taking care of laboratory pigs has the duty to acquaint themselves with the biology of the species and best serve the needs of the pig within the constraints of the experimental protocol under which the pig is being kept. Within rodent housing, a movement has begun, developing new housing facilities taking the natural environment of the animals into account[64] and we call for similar approaches in the future development of housing for laboratory pigs.

It is likely that the popularity of the pig as a laboratory animal will continue to increase, and the potential applications of the pig will expand. At present, there is, however, still a paucity of scientific literature available on the laboratory pig with respect to housing, management, handling, and training – all of crucial importance for the welfare of the animals, the handlers, and the quality of the scientific outcomes. Understanding the cognitive abilities, behavioral priorities, and emotions of laboratory pigs lies at the very heart of improving their welfare – and in contrast to pigs kept for meat, where keepers have no time and are only there for minutes each day – prioritizing cognitive enrichment in laboratory pigs is within reach.[65]

The present knowledge about the cognitive abilities of pigs calls for further focus on this type of enrichment to improve the welfare of pigs kept for research purposes,[66] especially under conditions demanding sterility or intensive housing, such as metabolism chambers, where the use of environmental enrichment is challenged. Through cognitive interaction with the environment or humans, the animals can keep a certain level of control over the environment, and essential resources, such as food or water, can be used as rewards for successful coping or as reinforcers in training setups.[67,68] Cognitive enrichment of pigs provided as positive reinforcement training or other rewarding tasks should be implemented in the daily management to the highest possible extent.

In addition, the complex cognitive abilities of pigs mean that future caretakers may take advantage of the ability of pigs to plan and anticipate future events. Pigs have been suggested to be able to predict potential challenges[69] and plan appropriate actions,[18] which also render them vulnerable to anxiety about the future, and memory of unpleasant events.[70] For example, pigs seem to be able to learn associations between visual cues and the duration of a subsequent period of confinement.[71] These abilities could be utilized by, e.g., signaling the duration of upcoming husbandry procedures, hence increasing the predictability and thereby reduce the aversiveness. This would also prompt the caregivers to consider with increased awareness and empathy how the pigs perceive the human-animal interactions – and how to promote meaningful, positive interactions with the animals.

REFERENCES

1. Roth, J. A. & Tuggle, C. K. Livestock models in translational medicine. *ILAR Journal* **56**, 1–6 (2015).
2. McCrackin, M. A. & Swindle, M. M. Biology, handling, husbandry, and anatomy. In Swindle, M. M. & Smith, A. C. (Eds.), *Swine in the Laboratory* (3rd edition) (Boca Raton, Florida: CRC Press, 2016), pp. 1–38.
3. Jensen, P. The weaning process of free-ranging domestic pigs – within-litter and between-litter variations. *Ethology* **100**, 14–25 (1995).
4. D'Eath, R. B. & Turner, S. P. 2009. The natural behavior of the pig. In Marchant-Forde, J. N. (Ed.), *The Welfare of Pigs* (Dordrecht, the Netherlands: Springer, 2009), pp. 13–46.
5. Herskin, M. S. & Jensen, K. H. Effects of different degrees of social isolation on the behaviour of weaned piglets kept for experimental purposes. *Animal Welfare* **9**, 237–249 (2000).
6. Ruis, M. A. W. et al. Adaptation to social isolation: Acute and long-term stress responses of growing gilts with different coping characteristics. *Physiology & Behavior* **73**, 541–551 (2001).
7. Jensen, P. Nest building in domestic sows: The role of external stimuli. *Animal Behavior* **45**, 351–358 (1993).
8. Pajor, E. A. Sow housing: Science, behavior, and values. *JAVMA* **226**, 1324–1344 (2005).
9. Hsia, L. C. & Wood-Gush, D. G. M. Social facilitation in the feeding behaviour of pigs and the effect of rank. *Applied Animal Ethology* **11**, 265–270 (1984).
10. Bollen, P. & Ritskes-Hoitinga, M. The welfare of pigs and minipigs. In Kaliste, E. (Ed.), *The Welfare of Laboratory Animals* (Dordrecht, the Netherlands: Springer, 2007), pp. 325–356.
11. Stewart, K. & Cabezon, F. Heat stress physiology in swine. Purdue University Extension Factsheet AS-362-W (2016). Available: https://www.extension.purdue.edu/extmedia/AS/AS-362-W.pdf
12. Held, S., Cooper, J. J. & Mendl, M. T. Advances in the study of cognition, behavioral priorities and emotions. In Marchant-Forde, J. N. (Ed.), *The Welfare of Pigs* (Dordrecht, the Netherlands: Springer, 2009), pp. 47–94.
13. Root-Bernstein, M., Narayan, T., Cornier, L. & Bourgeois A. Context-specific tool use by *Sus cebifrons*. *Mammalian Biology* **98**, 102–110 (2019).
14. Croney, C. C., Adams, K. M., Washington, C.G. & Stricklin, W. R. A note on visual, olfactory and spatial cue use in foraging behavior of pigs: Indirectly assessing cognitive abilities. *Applied Animal Behaviour Science* **83**, 303–308 (2003).
15. Bolhuis, J. E. et al. Working and reference memory of pigs (*Sus scrofa* domesticus) in a holeboard spatial discrimination task: The influence of environmental enrichment. *Animal Cognition* **16**, 845–850 (2013).
16. Broom, D. M., Sena, H. & Moynihan, K. L. Pigs learn what a mirror image represents and use it to obtain information. *Animal Behaviour* **78**, 1037–1041 (2009).
17. Held, S., Mendl, M., Devereux, C. & Byrne, R. W. Foraging pigs alter their behaviour in response to exploitation. *Animal Behaviour* **64**, 157–165 (2002).
18. Held, S. et al. Domestic pigs, *Sus scrofa*, adjust their foraging behaviour to whom they are foraging with. *Animal Behaviour* 79, 857–862 (2010).

19. Mendl, M., Held, S. & Byrne, R. W. 2010. Pig cognition. *Current Biology* **20**, R796–R798 (2018).
20. Kristensen, H. H., Jones, R. B., Schofield, C. P., White, R. P. & Wathes, C. M. The use of olfactory and other cues for social recognition by juvenile pigs. *Applied Animal Behaviour Science* **72**, 321–333 (2001).
21. Illmann, G., Schrader, L., Pinka, M. & Ustr, P. Acoustical mother-offspring recognition in pigs (*Sus scrofa domesticus*). *Behaviour* **139**, 487–505 (2002).
22. Koba, Y. & Tanida, H. How do miniature pigs discriminate between people? Discrimination between people wearing coveralls of the same colour. *Applied Animal Behaviour Science* **73**, 45–58 (2001).
23. Nawroth, C., Ebersbach, M. & von Borell, E. Are juvenile domestic pigs (*Sus scrofa domesticus*) sensitive to the attentive states of humans? – The impact of impulsivity on choice behavior. *Behavioural Processes* **96**, 53–58 (2013).
24. Nawroth, C., Ebersbach, M. & von Borell, E. Juvenile domestic pigs (*Sus scrofa domesticus*) use human-given cues in an object choice task. *Animal Cognition* **17**, 701–713 (2013).
25. Sommer, V., Lowe, A. & Dietrich, T. Not eating like a pig: European wild boar wash their food. *Animal Cognition* **19**, 245–249 (2016).
26. Murphy, E., Nordquist, R. & van der Staay, F. J. A review of behavioral methods to study emotion and mood in pigs, *Sus scrofa. Applied Animal Behaviour Science* **159**, 9–28 (2014).
27. Reimert, I., Bolhuis, J. E., Kemp, B. & Rodenburg, T. B. Indicators of positive and negative emotions and emotional contagion in pigs. *Physiol. Behav.* **109**, 42–50 (2013).
28. Goumon, S. & Spinka, M. Emotional contagion of distress in young pigs is potentiated by previous exposure to the same stressor. *Animal Cognition* **19**, 501–511 (2016).
29. Mendl, M., Burman, O. H. P., Parker, R. M. A. & Paul, E. S. Cognitive bias as an indicator of animal emotion and welfare: Emerging evidence and underlying mechanisms. *Applied Animal Behaviour Science* **118**, 161–181 (2009).
30. Douglas, C., Bateson, M., Walsh, C., Bédué, A. & Edwards, S. A. Environmental enrichment induces optimistic cognitive biases in pigs. *Applied Animal Behaviour Science* **139**, 65–73 (2012).
31. Bradbury, A. G., Eddleston, M. & Clutton, R. E. Pain management in pigs undergoing experimental surgery; a literature review (2012–2014). *British Journal of Anaesthesia* **116**, 37–45 (2016).
32. Castel, D., Willentz, E., Doron, O., Brenner, O. & Meilin, S. Characterization of a porcine model of post-operative pain. *European Journal of Pain* **18**, 496–505 (2014).
33. Pairis-Garcia, M. D. et al. Behavioural evaluation of analgesic efficacy for pain mitigation in lame sows. *Anim. Welfare* **24**, 93–99 (2015).
34. Di Giminiani P, et al. The assessment of facial expressions in piglets undergoing tail docking and castration: Toward the development of the piglet grimace scale. Frontiers in Veterinary Science 3, 1–10 (2016).
35. Viscardi, A. V. & Turner, P. V. Efficacy of buprenorphine for management of surgical castration pain in piglets. *BMC Veterinary Research* **14**, 318 (2018).
36. Herskin, M. S. & di Giminiani, P. Pain in pigs: Characterisation, mechanisms and indicators. In Spinka, M. (Ed.), *Advances in Pig Welfare* (Duxford, UK: Woodhead Publishing, 2018), pp. 325–356.
37. Richter, S. H., Garner, J. P. & Würbel, H. Environmental standardization: Cure or cause of poor reproducibility in animal experiments? *Nature Methods* **6**, 257–261 (2009).
38. Richter, S. H., Garner, J. P., Auer, C., Kunert, J. & Würbel, H. Systematic variation improves reproducibility of animal experiments? *Nature Methods* **7**, 167–168 (2010).
39. Brown-Brandl, T. M. et al. Heat and moisture production of moern swine. *ASHRAE* Transactions **NY-14-043**, 469–489 (2014).
40. Cabezon, F. A. et al. Technical note: Initial evaluation of floor cooling on lactating sows under acute heat stress. *The Professional Animal Scientist* **33**, 254–260 (2017).
41. Vasdal, G., Mogedal, I., Bøe, K. E., Kirkden, R. & Andersen, I. L. Piglet preference for infrared temperature and flooring. *Applied Animal Behaviour Science* **122**, 92–97 (2010).
42. Herskin, M. S., Bonde, M. K, Jorgensen, E. & Jensen, K. H. Decubital shoulder ulcers in sows: A review of classification, pain and welfare consequences. *Animal* **5**, 757–766 (2011).
43. Jensen, T. B. & Toft, N. Causes of and predisposing risk factors for leg disorders in growing-finishing pigs. *CAB Reviews: Perspectives in Agriculture, Veterinary Science, Nutrition and Natural Resources* **4**, 1–8 (2009).
44. Mouttotou, N., Hatchell, F. M. & Green, L. E. The prevalence and risk factors associated with forelimb skin abrasions and sole bruising in preweaning piglets. *Preventive Veterinary Medicine* **39**, 231–245 (1999).
45. Herskin, M. S. et al. Impact of the amount of straw provided to pigs kept in intensive production conditions on the occurrence and severity of gastric ulceration at slaughter. *Research in Veterinary Science* **104**, 200–206 (2016).

46. Newberry, R. C. Environmental enrichment: Increasing the biological relevance of captive environments. *Applied Animal Behaviour Science* **44**, 229–243 (1995).

47. Van de Weerd, H. A., Docking, C. M., Day, J. E. L., Avery, P.J. & Edwards, S. A. A systematic approach towards developing environmental enrichment for pigs. *Applied Animal Behaviour Science* **84**, 101–118 (2003).

48. Studnitz, M., Jensen, M. B. & Pedersen, L. J. Why do pigs root? A review on the need of pigs for foraging and exploration. *Applied Animal Behaviour Science* 107, 183–197 (2007).

49. Bracke, M. B. M. Chains as proper enrichment for intensively-farmed pigs? In Spinka, M. (Ed.), *Advances in Pig Welfare* (Woodhead Publishing, 2018), pp. 167–197.

50. Smith, M. E., Gopee, N. V. & Ferguson, S. A. Preferences of minipigs for environmental enrichment objects. *Journal of the American Association for Laboratory Animal* **48**, 391–394 (2009).

51. Marchant-Forde, J. N. & Herskin, M. S. Pigs as laboratory animals. In Spinka, M. (Ed.), *Advances in Pig Welfare* (Woodhead Publishing, 2018), pp. 445–476.

52. Casey, B., Abney, D. & Skoumbordis, E. A playroom as a novel swine enrichment. *Laboratory Animals* **36**, 32–34 (2007).

53. De Briyne, N., Berg, C., Blaha, T., Palzer, A. & Temple, D. Phasing out pig tail docking in the EU – present state, challenges and possibilities. *Porcine Health Management* **4**, 27 (2018).

54. Proudfoot, K. & Habing, G. Social stress as a cause of diseases in farm animals: Current knowledge and future directions. *The Veterinary Journal* **206**, 15–21 (2015).

55. Marchant-Forde, J. N. & Marchant-Forde, R. M. Methods to reduce aggression at mixing in swine. *Pig News & Information* **26** (2005), 63N–73N.

56. Rault, J. L. Friends with benefits: Social support and its relevance for farm animal welfare.*Applied Animal Behaviour Science* **136**, 1–14 (2012).

57. Herskin, M. S. & Hedemann, M. S. Effects of surgical catheterization and degree of isolation on the behavior and exocrine pancreatic secretion of newly weaned pigs. *Journal of Animal Science* **79**, 1179–1188 (2001).

58. DeBoer, S. P. et al. Does the presence of a human effect the preference of enrichment items in young isolated pigs? *Applied Animal Behaviour Science* **143**, 96–103 (2013).

59. DeBoer, S. P. et al. An initial investigation into the effects of social isolation and enrichment on the welfare of laboratory pigs housed in the PigTurn System assessed using tear staining, behaviour, physiology and haematology. *Animal Welfare* **24**, 15–27 (2015).

60. McGlone J, et al. Shipping stress and social-status effects on pig performance, plasm-cortisol, natural-killer-cell activity and leukocyte numbers. *Journal of Animal Science* **71**, 888–896 (1993).

61. Dalin, A., Magnusson, U., Haggendal, J. & Nyberg, L. The effect of transport stress on plasma levels of catecholamines, cortisol, corticosteroid-binding globulin, blood-cell count, and lymphocyte-proliferation in pigs. *Acta Veterinaria Scandinavica* **34**, 59–68 (1993).

62. Smith, A. C. & Swindle, M. M. Preparation of swine for the laboratory. *ILAR Journal* **47**, 358–363 (2006).

63. Rault J-L, et al. Gentle abdominal stroking ('belly rubbing') of pigs by a human reduces EEG total power and increases EEG frequencies. *Behavioural Brain Research* **374**, 111892 (2019).

64. Makowska, I. J. & Weary, D. M. The importance of burrowing, climbing and standing upright for laboratory rats. *Royal Society Open Science* **3**, 160136 (2016).

65. Mendl, M. T. & Paul, E. S. 2019. Assessing affective states in animals. In McMillan, F. (Ed.), *Mental Health and Well-being in Animals*. (2nd edition) (CABI Publishing, 2019).

66. Zebunke, M., Langbein, J., Manteuffel, G. & Puppe, B. Autonomic reactions indicating positive affect during acoustic reward learning in domestic pigs. *Animal Behaviour* **81**, 481–489 (2011).

67. Zebunke, M., Puppe, B. & Langbein, J. Effects of cognitive enrichment on behavioral and physiological reactions of pigs. *Physiology & Behavior* **118**, 70–79 (2013).

68. Sørensen, D. B. Never wrestle with a pig … *Laboratory Animals* **44**, 159–161 (2010).

69. Imfeld-Mueller, S., Van Wezemaela, L., Stauffachera, M., Gygax, L. & Hillmann, E. Do pigs distinguish between situations of different emotional valences during anticipation? *Applied Animal Behaviour Science* **131**, 86–93 (2011).

70. Marino, L. & Colvin, C. M. Thinking pigs: A comparative review of cognition, emotion and personality in *Sus domesticus*. *International Journal of Comparative Psychology* **28**, 23859 (2015).

71. Spinka, M., Duncan, I. J. H., & Widowski, T. M. Do domestic pigs prefer short-term to medium-term confinement? *Applied Animal Behaviour Science* **58**, 221–232 (1998).

Index

Printed in the United States
By Bookmasters